Assessing the Ability of the Afghan Ministry of Interior Affairs to Support the Afghan Local Police

Jefferson P. Marquis, Sean Duggan, Brian J. Gordon, Lisa Miyashiro

Prepared for the Office of the Secretary of Defense

For more information on this publication, visit www.rand.org/t/RR1399

Library of Congress Cataloging-in-Publication Data is available for this publication.
ISBN: 978-0-8330-9450-6

Published by the RAND Corporation, Santa Monica, Calif.

Cover: U.S. Army photo by Spc. Jessica Reyna DeBooy/Released.

The RAND Corporation is a research organization that develops solutions to public policy challenges to help make communities throughout the world safer and more secure, healthier and more prosperous. RAND is nonprofit, nonpartisan, and committed to the public interest.

RAND's publications do not necessarily reflect the opinions of its research clients and sponsors.

Preface

The RAND Corporation was asked to assess the ability of the Afghan Ministry of Interior Affairs to support the Afghan Local Police (ALP) program. To this end, this report discusses a range of logistics, personnel management, and training activities essential to the success of ALP's local security mission, as well as the ways that the Special Operations Joint Task Force–Afghanistan/NATO Special Operations Component Command–Afghanistan and other Coalition trainers and advisers attempted to improve the ministry's chances of sustaining the ALP program. The report also identifies lessons from the ALP support mission that might prove useful when undertaking similar efforts to help build local security forces in the future.

This research was sponsored by the Special Operations Joint Task Force–Afghanistan/NATO Special Operations Component Command–Afghanistan. It was conducted within the International Security and Defense Policy Center of the RAND National Defense Research Institute, a federally funded research and development center sponsored by the Office of the Secretary of Defense, the Joint Staff, the Unified Combatant Commands, the Navy, the Marine Corps, the defense agencies, and the defense Intelligence Community.

For more information on the International Security and Defense Policy Center, see www.rand.org/nsrd/ndri/centers/isdp or contact the director (contact information is provided on web page).

Contents

Figures and Table

Figures

Table

Summary

To better capture lessons learned from building local security force capacity in Afghanistan, RAND Corporation researchers evaluated a range of logistics, personnel management, and training activities essential to the success of the Afghan Local Police (ALP) local security mission. Initially established by U.S. and North Atlantic Treaty Organization (NATO) special operations forces (SOF), the ALP program was designed to become part of the Afghan Ministry of Interior Affairs (MOI). While continuing to be funded and advised separately from the rest of MOI's forces, the program was established in 2010 as a component of the Afghan Uniform Police (AUP)—the civil law enforcement pillar of the Afghan National Police (ANP)—which took responsibility for training, equipping, manning, and supporting ALP guardians in villages throughout the country. From ALP's inception, Coalition support to MOI has proven critical in the development of the ALP program. The three major Coalition organizations then supporting ALP were the Special Operations Joint Task Force–Afghanistan/NATO Special Operations Component Command–Afghanistan (SOJTF-A/NSOCC-A), the National Training Mission–Afghanistan (NTM-A), and the Combined Security Transition Command–Afghanistan (CSTC-A). These organizations provided direct materiel support and training to ALP units in the field and indirect support through mentoring and advisory efforts at the ministerial level.

This report highlights and synthesizes various indications of MOI capability and capacity with respect to ALP. Additionally, it presents lessons from the experience of MOI's support to ALP—and the Coalition's efforts to assist in this endeavor—that might prove useful to the United States and its allies as they attempt to help build local security force capacity in other parts of the world.

This analysis relies primarily on interviews conducted in Afghanistan in February 2013 with 58 Coalition and Afghan officials involved in the ALP program at the national, regional, and local levels; at the time, the transition to full Afghan management of support to ALP was still under way. To a lesser extent, the analysis draws on quantitative logistics and personnel reporting provided by SOJTF-A/NSOCC-A from January 2012 to January 2013. In the second phase of our research, we supplemented data from these sources with additional interviews with nine Coalition officials in February 2015, as well as information collected from secondary reports, in an attempt to understand the extent to which ALP's support situation had changed after MOI's complete assumption of responsibility for the program in 2014. However, the Coalition's limited ability to independently observe and verify ALP-related activities in the field since the end of its involvement in major combat operations meant that it was not possible to conduct a detailed comparison of the results of MOI support efforts in 2013 and 2015.

Cross-Cutting Themes

We identified the following key findings from the two phases of our research effort.

Demonstrated MOI Support to ALP

Although MOI is a long way from being able to provide comprehensive support to ALP, the RAND study team found many examples in 2013 and 2015 of MOI's ability to manage ALP logistics, personnel, and training. The regional logistics centers allow provincial AUP officials to participate in Afghan-led logistics workshops and coordinate with national officials on a range of equipment, resupply, and maintenance issues affecting ALP. MOI officials from ALP headquarters conduct staff assistance visits to provincial and district headquarters, where they conduct audits of ALP records and inventories and hold wide-ranging summits with local officials, notables, and ALP guardians. In newly validated ALP districts, Afghans are now handling recruiting, vetting, and in-processing functions—albeit not always in accordance with ALP's establishment procedures. Likewise, there are signs of deterioration in the pay distribution system, but many ALP guardians are getting paid on a regular basis, and the spread of MOI's electronic payment system to ALP promises to reduce incidences of theft and corruption. And regional training centers (RTCs) appear to be churning out a sufficient number ALP graduates trained by AUP instructors to keep up with current growth and attrition in the force.

Unclear Capability Gaps

Despite these capability advances, MOI faces serious gaps in its ability to sustain the ALP program; however, the extent of these gaps is not fully understood. The Coalition decided to "take off the training wheels" with respect to ALP equipping, resupply, and other support functions at the same time that it was rapidly reducing its presence in Afghanistan. So, there was a limited period to conduct a direct, district-level assessment of how the Afghans were performing their support responsibilities on their own and to help fix problems that arose. Even with their limited visibility of conditions in ALP villages, Coalition advisers acknowledge that MOI still has serious weaknesses, including the quantity and quality of personnel dedicated to ALP support; the mechanisms designed to track and account for ALP personnel, pay, and equipment; the transportation and storage of supplies and equipment; vehicle maintenance; checkpoint construction and upkeep; medical care; and communications between national and district headquarters. Lastly, although those we interviewed in 2015 indicated that the RTCs are mostly able to satisfy ALP's current initial training requirements, only a limited number of mobile training teams are available to provide more-advanced training to ALP guardians in their district locations.

Lack of MOI Control Over Local Police Chiefs

The lack of MOI control over those who command ALP is believed to be at the heart of many support problems. Because district and provincial chiefs of police work primarily for district and provincial governors, respectively, rather than for the MOI deputy minister of security, they are able to exercise considerable discretion over the provisioning of ALP and other ANP elements. Unless this organizational problem is fixed, they contend, ALP logistics will never function properly without continuing Coalition support.

Unclear MOI Commitment to ALP

Notwithstanding what is said in ALP establishment procedures about ALP program growth, duration, and support, MOI's commitment to ALP remains unclear. Some ALP headquarters officials we interviewed believe their ANP counterparts in other branches would never fully accept that ALP deserved their support. SOF advisers reported that MOI would drop ALP immediately if the U.S. government did not provide all of the resources for the ALP program and insisted that MOI use these resources to support ALP. The MOI officials we interviewed did not explicitly say this. Indeed, some praised the contributions of ALP and expressed a desire for more ALP resources. Yet they also expressed lingering concerns about the tribal balance within ALP units, as well as ALP guardian vetting and professionalism.

MOI commitment to ALP is even more complicated below the ministerial level. Although we did not have the opportunity to interview many Afghan officials in the field, our interviews with Coalition officials indicate that mid-level MOI support for ALP is uneven. Despite a groundswell of interest from many quarters in ALP startups, provincial and district officials have hoarded or diverted a significant amount of equipment and supplies intended for ALP guardians in villages. In addition, ALP officials voiced dissatisfaction with the Quick Reaction Force assistance they are supposed to receive from local ANP and Afghan National Army units.

Uncertainty in ALP's Future Mission

The final hurdle to judging MOI's ability to support ALP is the uncertainty in ALP's future mission. Several senior MOI officials we interviewed seemed interested in converting ALP into something resembling low-level AUP, who would be paid, employed, and deployed accordingly. Obviously, any such changes in strategy will affect ALP's sustainability. Fortunately, MOI is engaged with the international community in developing a comprehensive police reform plan. This should offer an opportunity to better understand, as well as shape, MOI's overall strategy with respect to all the elements of ANP, including the ALP program.

Logistics

Under Coalition pressure, MOI has made significant strides to improve its logistics practices and results. The Coalition's 2012 initiative to eliminate bottlenecks when providing ALP with initial equipment was mostly successful, and its chances for continued success under full Afghan control are relatively good, particularly if RTCs can continue to be used as the main venue for supplying new guardians with their initial kits.

Yet more needs to be done if the Afghans are to acquire a full and independent capacity to requisition, track, store, transport, distribute, and maintain necessary quantities of ALP equipment and supplies. In some cases, the Afghans' inability to master or follow the steps of MOI's requisition process has led to problems in ordering the equipment and supplies authorized for ALP. In other cases, the problem is district or provincial leaders' reluctance to provide ALP with authorized supplies, which stems from those leaders either having other uses in mind for those supplies or believing that the guardians will misuse them unless allotted sparingly. Many Coalition officials believe that the Western-oriented "pull" process for requesting supplies goes against the Afghan grain and that a Soviet-style "push" system is better suited to a security force with limited administrative capacity. In addition to resupply problems, ALP also faces logistical challenges related to the loss of Coalition air transport, limited provincial and district storage facilities, inadequate vehicle-maintenance education and support, limited medical capabilities and support, and unreliable communications between ALP headquarters and district headquarters.

Personnel Management

Afghans have begun to acquire and demonstrate many of the capabilities necessary to successfully manage ALP personnel. For example, Afghan elders, government officials, and contractors are already handling all ALP recruiting, vetting, and in-processing tasks with no assistance from the Coalition. Furthermore, MOI is, for the most part, paying ALP personnel in a timely manner. In addition, ALP guardians are starting to use electronic payment methods, which could lead to reduced corruption and delay.

Despite these hopeful signs, it is not yet clear whether the Afghans can maintain the current level of personnel management capability without Coalition mentorship and assistance. One indicator of managerial weakness is MOI's inability to authorize and fill many of the positions needed to support ALP guardians and other members of ANP. Also, some doubt MOI's ability to maintain tribal balances within ALP units, adequately vet ALP recruits, or conduct the full range of in-processing tasks in the future. For their part, MOI officials complain that ALP guardian pay is insufficient for the important work they do and is a source of tension between ALP and other elements of ANP.

Training

The state of ALP training is good compared with the situation in the rest of the Afghan police force. Approximately 86 percent of the ALP force has attended a formal training course, and 14 percent remain temporarily untrained, mostly because security threats prevent them from leaving their villages. Furthermore, the regional training concept introduced by the Coalition in late 2012 appears to be an essential component of the long-term training solution for ALP following the departure of the last remaining village-based Coalition SOF trainers and mentors in 2014. The RTCs selected for ALP training, many of which are colocated with regional logistics centers, provide a one-stop opportunity for ALP in-processing, equipping, and training. Although not without some start-up problems, the RTCs and their AUP trainers have demonstrated the ability to meet the demand for trained ALP recruits during a period when the force has faced significant attrition.

Nevertheless, there are still security and logistics concerns in transporting ALP guardians to training centers, and these concerns must be addressed by provincial and district chiefs of police, who may have other priorities. Also, guardians require follow-on and advanced training not offered to ALP personnel at the RTCs. Thus, most Coalition advisers we spoke to agreed that a hybrid training system—with local and regional aspects, and possibly the use of mobile training teams—was the best option for the future.

Lessons Learned

The following list of recommendations for future capacity-building missions derives from the RAND team's conversations with commanders and staff within SOJTF-A/NSOCC-A, NTM-A, CSTC-A, and MOI in 2013 and 2015. Although these recommendations may seem most pertinent to the circumstances of Afghanistan, we hope that they may also resonate with U.S. and allied government officials who are wrestling with how to help nations beset with insecurity, poor infrastructure, and weak governance develop and sustain local security forces without prolonged, direct external assistance.

Lesson 1: Advisers Must Take Account of the Operating Environment and Work in Concert with Various Partners

The obstacles confronting those charged with supporting ALP are formidable. As just indicated, they include the lack of MOI authority over provincial chiefs of police, the uncertain commitment of senior MOI leaders to the ALP program, and the difficulties in assessing conditions in ALP districts, among others. Although U.S. advisers are unlikely to confront the same array of issues in the future as they did in Afghanistan, they could face similar obstacles when assisting governments in poor, conflict-torn countries with limited infrastructure and human capital. In such cases, they must do their best to first understand the lay of the land and then recommend a support plan that either circumvents or erodes potential blockages. Furthermore, in order to lessen bureaucratic resistance, local security advisers need to collaborate on the support plan with officials in the host nation and in other U.S. and international assistance organizations. Although there were certain advantages that stemmed from the decision of U.S. and NATO SOF to build ALP from the ground up largely on its own, the proprietary nature of the ALP enterprise made it challenging to hand off the program to MOI as the Coalition began its withdrawal or to obtain the enthusiastic backing of Coalition agencies responsible for advising and assisting the rest of the Afghan police force.

Lesson 2: Pull-Based Logistics Systems Often Take a Long Time to Evolve

SOJTF-A/NSOCC-A and its Coalition partners have faced a quandary throughout the existence of the ALP program. Should they continue to push as much equipment and supplies as authorized to ALP districts (to avoid their being overrun by the Taliban), or should they insist that MOI provide an accurate accounting of existing inventories of ALP supplies before releasing additional resources (out of a suspicion that a significant portion would not reach their intended destination)? That this dilemma has never been satisfactorily resolved has much to do with the length of time it takes to establish an effective pull-based stock replenishment system. What this experience suggests is that in situations where the existing supply chain is long, broken, or undeveloped, U.S. advisers and partner government officials may have little choice initially but to push resources to units in the field to ensure that they have the wherewithal to defend themselves. As soon as possible, however, they should begin to put in place a supply system that allows for positive control through property accountability or inventory records so that equipment and supplies can be tracked and their organizational owners held responsible for their maintenance and use. Rather than attempt to make the immediate leap to a first-tier, pull-based stock replenishment system, donors should consider simpler alternatives that account for the partner's level of resources, literacy, technical competence, communications, and data availability. In addition, advisers and partner officials should look for creative ways to collect and analyze available information that might contribute to a greater understanding of logistics requirements and consumption patterns. In the longer term, they should invest in logistics-focused human capital development so that high-quality personnel can be recruited and trained, thereby setting the stage for a higher-functioning logistics operation.

Lesson 3: Managing Dispersed Forces Requires a Balance Between Local Autonomy and Central Oversight

Although MOI is beginning to develop effective processes for the administration of ALP personnel, managing the local security forces in widely dispersed locations has proved to be challenging. As a result, provincial and district police officials have been given a relatively free hand to support, neglect, or exploit ALP forces. What this indicates is the need to find a balance between encouraging local leaders to take charge of the daily management of local security forces and ensuring that the former raise and employ the latter appropriately and continue to provide adequate support to them. Although it is ultimately up to the host-nation government to determine appropriate managerial structures, policies, and processes, external advisers should do what they can to help host-nation leaders overcome handicaps to effective oversight. For example, they could advise the national headquarters to improve its downward lines of communication and influence by placing trusted agents at regional logistics centers and local administrative headquarters. Beyond that, they could back efforts to place local police officials under national ministerial authority in places where they currently are not, such as Afghanistan.

Lesson 4: Centralized Training Has Advantages, but a Hybrid System May Work Best Over the Long Term

RTCs in Afghanistan have proven successful so far in turning out a steady supply of ALP graduates, and they have improved the chances that Afghans can sustain the program over the long term. However, regional training cannot entirely substitute for the localized training that SOF teams provided in villages throughout much of Afghanistan from 2010 to 2014. Moreover, follow-on training at the district level is necessary to maintain and improve ALP skills, as well as to establish bonds of mutual support and trust between village guardians and other security forces in their vicinity. It also provides an opportunity for training providers to assess and monitor local security force performance in the field. With respect to future capacity-building missions, local security force advisers should consider recommending a hybrid (local and regional) training system. This would require two planning steps. The first is a comprehensive assessment of the training needs of all of the elements of the police force, as well as the capacity of RTCs to meet these needs. Second, based on the results of this assessment and a detailed evaluation of what is realistic and appropriate for particular elements of the local security force, external advisers and host-nation officials should develop training plans that employ a combination of RTCs, local training venues, and mobile training teams.

Lesson 5: The Coalition Advisory Structure Should Be Maintained Until the Host Nation Has an Assured Sustainment Capability

Despite promising developments in logistics, personnel management, and training, it is still unclear whether Afghanistan's MOI has the will and capability to independently support the ALP program without outside assistance and oversight. Thus, if the United States and its allies desire to help build local security forces in the future in countries similar to Afghanistan, where infrastructure is limited and governance is weak, they should consider maintaining an advisory presence in key regions, or some type of circuit rider or distributed operations scheme, in the period following the transition. This would permit coalition advisers to continue to work with headquarters officials and local leaders to resolve resource-, personnel-, and training-management issues pertaining to the police and military. This is not to say that outsiders should assume the host government's role of ensuring adequate support to local security forces. External advisers should focus on encouraging local officials to establish their own mechanisms for managing the support requirements of local security forces, as well as overseeing the implementation of these managerial functions.

As the largest U.S. military–sponsored experiment in building the capacity of village-based security forces since Vietnam, the ALP program provides a rich repository of lessons learned—both constructive and cautionary—for future security force assistance missions in conflicted and developing parts of the world. Of particular importance is the lesson to engage with the host nation early on about its willingness to independently support local security forces and, assuming it proves willing, map out the steps required to give it the capability to sustain these forces without direct foreign assistance.

Acknowledgments

The research for this report could not have been completed without considerable help from many individuals within the Special Operations Joint Task Force–Afghanistan/NATO Special Operations Component Command–Afghanistan (SOJTF-A/NSOCC-A), the National Training Mission–Afghanistan (NTM-A), and the Combined Security Transition Command–Afghanistan (CSTC-A), who not only took the time to talk to the RAND Corporation analysts conducting this study, but who also provided us with considerable guidance in navigating a complex and somewhat opaque topic, as well as with logistical support and access to a range of officials within the Afghan Ministry of Interior Affairs. We would like to offer special thanks to Maj Derek Williamson of SOJTF-A/NSOCC-A and LTC Gerald Faunce of NTM-A for their invaluable assistance in making arrangements for the RAND team's visit to Afghanistan in February 2013; to junior and senior officials within Afghanistan's Ministry of Interior Affairs who spoke to us at length in 2013 about their perspectives on the ALP program and the ministry's support to it; and to COL Paul Roberts, SGM Randall Krueger, and MAJ William Love for their assistance in February 2015.

The authors would also like to express their appreciation to Linda Robinson for her review of an earlier version of this report and to the reviewers of the current report, Eric Peltz and Mark Moyar. Their comments and recommended changes greatly improved the contents of this publication.

Abbreviations

ALP	Afghan Local Police
ANA	Afghan National Army
ANP	Afghan National Police
AUP	Afghan Uniform Police
CJSOTF-A	Combined Joint Special Operations Task Force–Afghanistan
CSTC-A	Combined Security Transition Command–Afghanistan
DCOP	district chief of police
GIRoA	Government of the Islamic Republic of Afghanistan
MOI	Afghan Ministry of Interior
NATO	North Atlantic Treaty Organization
NSOCC-A	NATO Special Operations Component Command–Afghanistan
NTM-A	National Training Mission-Afghanistan
PCOP	provincial chief of police
RLC	regional logistics center
RTC	regional training center
SOAG	special operations advisory group
SOF	special operations forces
SOJTF-A	Special Operations Joint Task Force–Afghanistan

CHAPTER ONE

Introduction

In winter 2013, the RAND Corporation was asked to assess the ability of the Afghan Ministry of Interior Affairs (MOI) to support the Afghan Local Police (ALP) program following the expected departure of Coalition combat forces at the end of 2014.[1] The intent was to evaluate the range of logistics, personnel management, and training activities essential to the success of ALP's local security mission and suggest ways that the three major Coalition organizations then supporting ALP—the Special Operations Joint Task Force–Afghanistan/NATO Special Operations Component Command–Afghanistan (SOJTF-A/NSOCC-A), the National Training Mission–Afghanistan (NTM-A), and the Combined Security Force Transition Command–Afghanistan (CSTC-A)—could best use their train, advise, and assist resources to improve MOI's chances of sustaining the ALP program in the future.[2]

[1] ALP was initially established by U.S. and North Atlantic Treaty Organization (NATO) special operations forces (SOF) in 2010.

[2] Established in 2012 and numbering roughly 13,000 special operators and support personnel from 25 nations at its peak, SOJTF-A/NSOCC-A is a novel division-level headquarters that encompasses all in-country NATO special operations assets and forces. Its missions span the entire spectrum of special operations, from direct action to capacity-building. The latter has entailed advising and assisting Afghan SOF, as well as standing up the ALP program and helping transition it to MOI control. Until the end of 2014, SOJTF-A/NSOCC-A worked closely with NTM-A and CSTC-A on ministerial-level advisory and support issues related to ALP. In 2015, NTM-A was stood down, and its relevant functions shifted to SOJTF-A/NSOCC-A and CSTC-A. The latter now has the primary role in the training and development of all Afghan security forces. See Karen Parrish, "Special Ops Task Force Helps Shift Afghanistan Trend Line," American Forces Press Service, May 13, 2013. From this point forward in the report, we refer to SOJTF-A/NSOCC-A as NSOCC-A, for brevity.

Two years later, in winter 2015, the RAND study team returned to Afghanistan to update its evaluation of MOI support to ALP and present lessons from MOI's experience—and the Coalition's assistance efforts—that might prove useful to the United States and its allies in their attempts to help build local security force capacity in other parts of the world. Thus, this report combines a detailed historical account of ALP's support situation, a recent (albeit limited) assessment of that situation, and some thoughts on grassroots security force assistance in difficult environments.

A Short History of ALP

To place our assessment of MOI support to ALP in context, this section provides background information on Afghan and Coalition efforts to make local defense forces into a major component of the counterinsurgency campaign in Afghanistan, which culminated in the establishment of the ALP program.

Precursors

Although efforts and proposals to establish Afghan self-defense at the local level had been going on almost since the American invasion in 2001,[3] by the beginning of 2009, there was a growing belief in Coalition and Afghan circles that a rural-based local security program would be required to defeat the Taliban. The perceived inefficacy of the Coalition's campaign in rural localities was believed to be a major contributor to the growth in the capability of the insurgency, and intelligence analyses suggested that insurgents were benefiting from a weak economy and ineffective governance outside major population centers. As a result, a pilot local defense initiative, the Afghan Public Protection

---

[3] This included ad hoc local defense forces (e.g., community watches, the Afghan Security Guard) and formal local defense forces (e.g., the Afghan National Auxiliary Police), as well as proposals for a comprehensive SOF-led local defense initiative (unpublished 2016 research on the legacy of SOF, by Daniel Egel and Stephanie Young). The rationale for local defense forces in counterinsurgency situations is explained in Seth G. Jones, *The Strategic Logic of Militia*, Santa Monica, Calif.: RAND Corporation, WR-913-SOCOM, 2012.

Program, was established in 2009. Designed to improve local security, deny insurgents support, and build the support and legitimacy of the Government of the Islamic Republic of Afghanistan (GIRoA),[4] the Afghan Public Protection Program never expanded beyond five districts in Wardak Province and was absorbed into the Afghan military in 2010. Nevertheless, it validated the local defense concept for NATO's International Security Assistance Force and paved the way for the Local Defense Initiative (initially known as the Community Defense Initiative) in July 2009. Also short-lived, the initiative was the first local defense force designed by U.S. and NATO SOF. Although endorsed by then–International Security Assistance Force Commander GEN Stanley McCrystal, the Local Defense Initiative lacked the support of the U.S. Embassy—which questioned MOI's ability to sustain the program once the Coalition had departed—and Afghan President Hamid Karzai, who was reportedly reluctant to buy into any program that was advocated by the Americans.[5]

Establishment and Mission

Established in 2010 by President Karzai as part of a bilateral agreement between the United States and GIRoA, the ALP program was more highly developed and longer lasting than previous local defense initiatives.[6] Like the Local Defense Initiative, ALP was created and enabled by U.S. and NATO SOF. In accordance with the new Village Stability Operations concept, SOF teams were embedded in villages, where they worked with locals to provide security, improve local gov-

---

[4] U.S. Department of Defense, *Report on Progress Toward Security and Stability in Afghanistan*, Washington, D.C., November 2010.

[5] Unpublished 2016 research on the legacy of SOF, by Daniel Egel and Stephanie Young.

[6] For early accounts of the development of ALP and its role in village stability operations, see Dan Madden, "The Evolution of Precision Counterinsurgency: A History of Village Stability Operations & the Afghan Local Police: Commander's Initiative Group," CFSOCC-A Commander's Initiative Group, June 30, 2011; Ali Nadir, "SOF, Shuras, and Shadow Governments: Informal Governance and Village Stability Operations," CFSOCC-A Commander's Initiative Group, 2011; and Lisa Saum-Manning, *VSO/ALP: Comparing Past and Current Challenges to Afghan Local Defense*, Santa Monica, Calif.: RAND Corporation, WR-936, 2012.

ernance, and support economic activity. Emphasis was placed equally on local legitimacy and connectivity with the Afghan government.[7] Local legitimacy was achieved through a robust local selection process in which ALP guardians were vetted by local councils and then trained and supported by the embedded SOF team. Connectivity to the government was ensured by placing the ALP program under the command of the local district chief of police (DCOP),[8] as well as by making the DCOPs and provincial chiefs of police (PCOPs) responsible for ALP pay and logistics.[9]

ALP's mission was encapsulated in two MOI documents: the *Ten-Year Vision for the Afghan National Police: 1392–1402* and the *ALP Policy and Establishment Procedures.*[10] The former stated MOI's vision for ALP: to "concentrate on ensuring security in insecure regions, to prevent infiltration of armed insurgents, and to provide the grounds for rehabilitation and economic development, so that the National Police can concentrate on providing community and civilian policing services."[11] The April 2013 *ALP Policy and Establishment Procedures* emphasized the community orientation of ALP, as well as its narrow security role. For example, it directed local authorities "not to move ALP members from one district to another district or area except in case of emergency."[12] Also, unlike other elements of the Afghan National Police (ANP), ALP guardians were not permitted to investigate crimes or make arrests.

---

[7] In this way, ALP differed from the two failed previous local defense initiatives—the Wardak-based Afghan Public Protection Program and larger Local Defense Initiative.

[8] ALP guardians would be recruited and vetted by villages across the district, in groups of 10–30 per village, to establish a network of mutually reinforcing forces.

[9] Unpublished 2016 research on the legacy of SOF, by Daniel Egel and Stephanie Young.

[10] Islamic Republic of Afghanistan, *Ten-Year Vision for the Afghan National Police: 1392–1402*, Kabul, Afghanistan: Ministry of Interior Affairs, 2013a; and Islamic Republic of Afghanistan, *ALP Policy and Establishment Procedures*, Kabul, Afghanistan: Ministry of Interior Affairs, April 2013b.

[11] Islamic Republic of Afghanistan, 2013a, p. 12.

[12] Islamic Republic of Afghanistan, 2013b.

Size and Growth

After growing rapidly in the program's first several years, the total number of ALP guardians stood at more than 22,000 in early 2013.[13] The security and local knowledge they provided led NSOCC-A to propose an expansion of the program to as high as 45,000 personnel, to include a commensurate increase in the personnel and equipment authorization for ALP, which is part of the *tashkil*.[14] This demonstration of support from the Coalition, as well as evidence of ALP success on the ground, convinced the Minister of Interior Affairs at the time to request an expansion of ALP tashkil authorizations from 30,000 to 45,000 guardians. This increase in forces was projected to cost $90 million per year more than the $180 million then allotted by the U.S. government, which was and remains the ALP program's sole source of funding. By 2015, however, concern over GIRoA's ability to support and sustain an expanded ALP force and continued wariness of ALP in the Afghan political establishment caused a change in policy regarding the intended size of the program. The number of guardians stood at 28,450 in March 2015,[15] and a Coalition official told RAND interviewers that there were no plans to raise or lower the 30,000-force ceiling authorized for the ALP by the President, which was considered appropriate and manageable for the MOI.[16]

ALP Organization and Command and Control

Managing ALP has been a challenge in part because of the exceptional way it has been incorporated into the national police force. Although the presidential decree establishing ALP created an administrative headquarters for the program, it specified that the Afghan Uniform Police (AUP), the civil law enforcement pillar of ANP, would act as the program's executive agent. AUP responsibilities include ensuring

---

[13] NSOCC-A, "ALP Weekly Update," Kabul, Afghanistan, April 22, 2013g.

[14] The *tashkil* is an organizational document that prescribes, among other things, the personnel and equipment authorized for Afghan military and police organizations.

[15] Data provided to RAND by the ALP Special Operations Advisory Group G1 Adviser, March 5, 2015, not available to the general public.

[16] NSOCC-A official, interview with the authors, February 25, 2015.

that the ALP program is properly managed, ALP sites are correctly established, and ALP forces are adequately trained, equipped, manned, and supported.[17] Thus, every official that either commands or supports ALP above the district leader, including those at ALP headquarters in Kabul, belongs to AUP, not ALP.[18]

Because of AUP difficulties in supporting, commanding, and overseeing ALP, some in the Coalition recommended separating the program from AUP and making it an independent pillar of MOI, with a direct line of authority from the national level to the district forces.[19] However, in 2012, NTM-A took issue with this course of action for several reasons. First, it argued, the drawdown of Coalition forces would leave insufficient partnering resources to create a fully functional, independent organization. Second, a separate pillar would drastically increase the requirement for ALP support personnel and would mean reducing other organizations' tashkils. Third, the establishment of a separate pillar would require additional supplies, personnel, and infrastructure for which MOI had not budgeted.[20] This argument eventually held sway.

In a bid to institutionalize the ALP without making it a standalone organization, the then–Deputy Minister of Interior Affairs announced in November 2012 that ALP would be permanently designated a component of AUP.[21] As a result, ALP became a subpillar of MOI. Although the program has a separate director and supporting staff that fall under the MOI Deputy Ministry for Security Affairs, the command and control of ALP guardians is the responsibility of AUP officials; in particular, the provincial and district chiefs of police.[22] In addition, the DCOP in ALP districts is officially the ALP command-

---

[17] Deputy Commander of Special Operations Forces, NTM-A, "Info Paper 12-05," Kabul, Afghanistan, 2012b.

[18] NSOCC-A official, interview with the authors, February 25, 2015.

[19] Deputy Commander of Special Operations Forces, 2012b.

[20] Deputy Commander of Special Operations Forces, 2012b.

[21] NSOCC-A, "ALP Weekly Update," Kabul, Afghanistan, April 8, 2013e.

[22] Deputy Commander of Special Operations Forces, NTM-A, "Primer: Mission Analysis/Needs Analysis," Kabul, Afghanistan, 2012c.

er.[23] ALP headquarters in Kabul acts as an advocate for ALP interests within MOI, but the headquarters has no authority to command or control ALP guardians in the field.[24]

Coalition Force Support Organization

Coalition support to MOI has proven critical in the development of the ALP program. NSOCC-A, NTM-A, and CSTC-A have provided direct materiel support and training to ALP units in the field and indirect support through mentoring and advisory efforts at the ministerial level. Until 2014, NSOCC-A, through its various components, advised and mentored ALP and AUP operational and support personnel from the village level to the provincial level.[25] Initially, NTM-A assumed the responsibility for managing the resources needed to equip ALP and providing advice and assistance to the ministerial-level officials who supported ALP.[26] Following the dissolution of NTM-A at the end of 2014, NSOCC-A took over the responsibility for advising and mentoring ALP headquarters through its ALP special operations advisory group (SOAG),[27] with CSTC-A handling the overall advisory relationship with MOI. Although the ALP SOAG has no direct role in supporting guardians in the districts, it continues to be involved in helping ALP headquarters resolve local ammunition, food, and water

---

[23] NSOCC-A, 2013e.

[24] NSOCC-A official, interview with the authors, February 25, 2015.

[25] NSOCC-A official, interview with Todd Helmus and Phillip Padilla, January 2013.

[26] U.S. Department of Defense Inspector General, *Special Plans and Operations: Assessment of the U.S. Government and Coalition Efforts to Develop the Afghan Local Police*, Alexandria, Va., July 9, 2012.

[27] Senior SOF commanders in Afghanistan created SOAGs to train, advise, and assist headquarters elements of the Afghan Special Security Forces. SOAGs are small formations of advisers who operate under NSOCC-A. Each SOAG is aligned with a headquarters element of the Afghan Special Security Forces, including ALP, General Command of Police Special Units, Commandos, Ktah Khas, and the Special Mission Wing. One SOAG works at both the Ministries of Defense and Interior Affairs and is referred to as the Ministry Advisory Group Special Operations Forces Liaison Element. See Todd C. Helmus, *Advising the Command: Best Practices from the Special Operations Advisory Experience in Afghanistan*, Santa Monica, Calif.: RAND Corporation, RR-949-OSD, 2015.

problems.[28] That said, with limited visibility on developments outside the capital, it is often impossible to know what is actually needed at the district level.[29]

ALP Special Operations Advisory Group Focus Areas

In 2015, the ALP SOAG identified the following four areas in which it planned to focus its advisory efforts, aimed at institutionalizing ALP and resolving certain issues regarding the future of the program:[30]

1. *ALP transition planning*: Considering a cost of $120 million for 30,000 guardians, ALP SOAG personnel tout ALP as the best "bang for the buck" in the Afghan security structure.[31] Although the program's funding seems secure through 2018, GIRoA's long-term plan for ALP is less certain. To assist with development of this plan, NSOCC-A has helped to build a consensus among Afghan officials, potential international donors, and village elders that ALP is a responsible security force with institutional backing and significant impact on local security.[32]
2. *Financial transparency*: ALP SOAG personnel have engaged their MOI counterparts to ensure accurate personnel-strength reporting, a key component of financial transparency. As discussed previously, the SOAG has limited visibility on local conditions beyond reports that ALP headquarters receives from local units.[33] Nonetheless, personnel reports are checked regularly, and ALP headquarters personnel are encouraged to conduct audits when visiting districts or conducting staff assistance visits. Another initiative to improve financial transparency is to

[28] NSOCC-A official, interview with the authors, February 25, 2015.

[29] NSOCC-A official, interview with the authors, February 25, 2015.

[30] SOJTF-A, "ALP Special Operations Advisory Group," policy memorandum, Kabul, Afghanistan, February 26, 2015.

[31] NSOCC-A official, interview with the authors, February 25, 2015.

[32] NSOCC-A official, interview with the authors, February 25, 2015.

[33] NSOCC-A official, interview with the authors, February 25, 2015.

ensure that guardians are being correctly paid. Since the inception of the ALP program, payments in many districts have been handled through a system of trusted agents (see Chapter Four). Coalition officials have pushed for electronic fund transfers to be established across the force. As of May 2015, 42 percent of ALP personnel were on such transfers.

3. *ALP survivability*: ALP personnel represent a significant portion of Afghan security force casualties, a result of their consistent presence at key points around their villages. To keep casualty rates as low as possible, Coalition advisers have pressed for the procurement and fielding of protective equipment and weapons to all ALP guardians.[34] As of February 2015, more than 7,000 AK-47 assault rifles were needed to realize this goal, and the SOAG was working to have the weapons either transferred from the Ministry of Defense or donated by a third country. Although some Coalition and MOI officials believe that a more advanced weapon should be fielded, ALP SOAG advisers insist that the AK-47 is the right weapon for ALP and that the guardians are comfortable with it and can be trained better on it than other on other types of rifles.[35]
4. *Oversight responsibilities of ALP headquarters*: Finally, the ALP SOAG has focused on the unique position of ALP headquarters within MOI. Despite its lack of operational responsibility for ALP units, Coalition mentors have tried to promote ALP head-

[34] From 2010 to 2012, 353 ALP personnel were killed in action, giving the program a casualty rate more than twice that of other security forces (Joris Fioriti, "Local Police, an Uncertain Player in Afghan Future," Goshta, Afghanistan: Agence France-Presse, December 31, 2012; and Mark Moyar, *Village Stability Operations and the Afghan Local Police*, Joint Special Operations University, October 2104, p. 47). If anything, the ALP casualties seem to have gotten worse, in both absolute and relative terms, since 2012 and, particularly, since the withdrawal of Coalition combat forces in 2014. According to the International Crisis Group, about 700 ALP personnel were killed and 800 injured in the Afghan calendar year from April 2013 to April 2014; and in the first three quarters of the 2014–2015 calendar year, the force suffered 1,015 killed and 1,320 injured, meaning that ALP casualties more than doubled (International Crisis Group, *The Future of the Afghan Police*, Brussels, Belgium, Asia Report No. 268, June 4, 2015).

[35] NSOCC-A official, interview with the authors, February 26, 2015.

> quarters' effectiveness by encouraging its commander and staff members to conduct staff assistance visits and battlefield circulations around Afghanistan. These visits potentially serve several purposes. First, they provide visibility on the state of ALP for those who are supposed to be the force's advocates within MOI. Second, the visits provide an opportunity to assess the degree to which ALP units meet personnel, financial, and logistics standards. Finally, the visits allow ALP leadership to meet and communicate with regional, provincial, and district officials who oversee the operations of ALP units.

Motivation for the Study

Although not without its critics, ALP has become a key component of the Coalition's counterinsurgency strategy in Afghanistan. This is particularly true in rural areas, where ANP and Afghan National Army (ANA) presence is relatively limited and the insurgency has traditionally been strong. Rhetorically at least, GIRoA and MOI have moved from a position of reluctant support for ALP to a position of acceptance of the program as a necessary security tool. This shift in attitudes is attributable to ALP's service as the first line of defense for the Afghan National Security Forces in many Taliban-infested districts—an undertaking that has caused them to suffer a disproportionate share of enemy attacks and attendant casualties. Nonetheless, MOI has not always considered ALP to be worthy of its sustained attention and support. Rather, many Afghans have seen ALP as a creation of the U.S. SOF that the Americans were expected to oversee and nourish. Coupled with a general lack of confidence in MOI's sustainment capability,[36] this situation has led some Coalition and ALP officials to express doubts about MOI's ability to support ALP, particularly during a time of diminishing Coalition influence.

Recognizing the challenge, NSOCC-A, NTM-A, and CSTC-A have worked closely with MOI to improve its support capacity. For

[36] NTM-A official, interview with the authors, February 17, 2013.

example, they have attempted to institutionalize the ALP program by enacting revised establishment procedures, to strengthen MOI's ability to independently train ALP recruits via regional training centers (RTCs), to increase the regularity of ALP pay through new payment mechanisms, and to increase MOI's logistics capacity by improving its equipping processes and its accountability and collaborative mechanisms. The open question was and remains whether there would be opportunity and time for these strategic initiatives to take hold as the country's security was being returned to Afghan control and the presence of Coalition trainers and advisers evaporated in rural areas.

However, there was not a lot of time for these strategic initiatives to take hold prior to transitioning nearly the full gamut of support responsibilities to the Afghans at the end of 2014. Since then, there are indications that certain support processes are functioning reasonably well from the perspective of U.S. advisers to MOI headquarters. But the extent of support provided to ALP at the provincial and district levels continues to vary considerably and is no longer possible to gauge in a systematic manner, given the Coalition's reliance on Afghan reporting mechanisms of uncertain reliability.

Study Approach

The study team relied primarily on a qualitative research approach, focused on an analysis of ALP program materials and interviews with ALP program stakeholders. With crucial assistance from NSOCC-A and NTM-A, members of the RAND team were able to interview 58 Coalition and Afghan officials at the national, regional, and local levels in February 2013. A majority of these interviews were conducted in Afghanistan's capital, Kabul. Those interviewed included the heads of the MOI directorates of relevance to ALP, along with their principal staff members and Coalition advisers. In 2013, the RAND team also attended a regional logistics conference in eastern Afghanistan; interviewed Coalition training and logistics mentors and advisers in Herat, in the western part of the country; and toured MOI's national warehouses located on the outskirts of the capital. In addition, we conducted a quantitative analysis of logistics and personnel data from

59 ALP districts provided by NSOCC-A for January 2012 to January 2013.[37]

We conducted follow-up interviews with nine members of the ALP SOAG and CSTC-A at the Resolute Command Headquarters in Kabul in February 2015 in order to document changes in the type and level of support and advisory assistance being provided to ALP since the RAND study team's previous visit to Afghanistan. (See Figure 1.1 for a breakdown of the Coalition and Afghan officials we interviewed.) Because of time, resource, and security constraints, we were unable to gather firsthand impressions of ALP conditions in the field during this trip or to speak with Afghan officials involved in managing or supporting the ALP program. However, we did supplement our research with recent reporting on ALP support issues provided by the Special Inspector General for Afghanistan Reconstruction and nongovernmental organizations. As a result, we were able to update our understanding of MOI logistics, personnel management, and training processes that affect ALP, but it was not possible to compile more than a general,

Figure 1.1
Coalition and Afghan Officials Interviewed in February 2013 and 2015

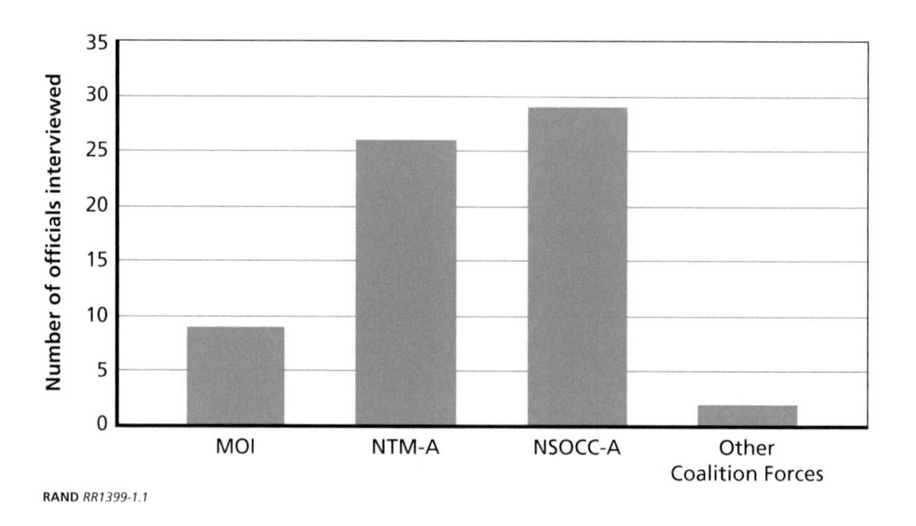

RAND *RR1399-1.1*

[37] NSOCC-A, ALP data provided to RAND, January 2012–January 2013.

Coalition-centric picture of MOI's progress in supporting ALP personnel in the diverse provinces and districts in which they reside.

The RAND team's research was guided by the following questions related to MOI's management of logistics, administrative, and training functions relevant to ALP:

1. To what extent can MOI independently provide logistics support to ALP? For example,
   a. initial equipment issue
   b. resupply
   c. maintenance support
   d. checkpoint facilities
   e. medical support
   f. communications support.
2. To what extent can MOI independently manage ALP personnel? For example,
   a. recruiting, vetting, and in-processing
   b. initial entry training
   c. ongoing training
   d. pay and benefits.
3. To what extent can MOI independently assess, monitor, and evaluate its managerial performance with respect to ALP and take necessary corrective action?
4. What are ALP's major logistical, administrative, and training challenges?
5. What steps have NSOCC-A and other Coalition training and advisory organizations taken to address these problems? How much progress has been made? What remains to be done? How cooperative has MOI been in carrying out these initiatives? What factors explain their willingness or lack of willingness to cooperate?
6. Does MOI have the funding and administrative capacity to support ALP currently? If not, what is being done to ensure that sufficient funding and administrative capacity will exist in the future?

In addition to analyzing documents and conducting interviews, the RAND team analyzed data from the district-level personnel and logistics tracking mechanism maintained by NSOCC-A for the period January 2012 to January 2013. A major goal of the team's quantitative research efforts was to estimate ALP support outcomes in "mature" tactical overwatch districts—that is, districts with ALP that had fully transitioned to Afghan control for at least several months, although they were still monitored by U.S. and NATO SOF teams.

Unfortunately, there is not enough information at this stage to make a confident prediction about ALP's institutionalization and sustainability. However, we are able to marshal a considerable body of evidence regarding MOI strengths and weaknesses in logistics, personnel management, and training that affect ALP (although the preponderance of our evidence was collected in 2013, prior to the complete turnover of the ALP program to MOI). Furthermore, we can document the steps that Afghan and Coalition officials have taken to overcome or reduce MOI deficiencies and can point out lessons they have learned in the past few years that might be applicable to future efforts of the United States and its allies to help stand up local security forces that are likely to retain the support of the host government.

Human subject protection protocols were used in this study and report in accordance with appropriate statutes and U.S. Department of Defense regulations. The views of sources rendered anonymous by these protocols are solely their own and do not represent the official policy or position of the Department, the U.S. government, or the government of Afghanistan.

Organization of the Report

This report is organized into six chapters. Chapter Two examines cross-cutting obstacles that affect all of the functional activities that MOI currently performs on behalf of ALP. These obstacles include MOI administrative capacity, MOI and ALP command and control, GIRoA's commitment to ALP, joint operational support to ALP, Coalition support for MOI and ALP, and ALP's future role in ANP.

The next three chapters examine specific problems in logistics, personnel management, and training. Chapter Three covers various logistics functions and describes the processes, initiatives, and results pertaining to them. These functions include equipment distribution and resupply, inventory management, transportation and storage, equipment maintenance, checkpoint facilities, and medical and communication support. Chapter Four focuses on personnel management issues affecting ALP, including authorizations and fill rates for ALP support positions; ALP recruiting, vetting, and in-processing procedures and results; and ALP pay processes, effectiveness, and sufficiency. Chapter Five discusses various training topics. These include the debate over the locus of ALP training, RTC planning and progress, and issues over who should train ALP and what ALP personnel should learn.

Chapter Six provides lessons learned from Coalition and Afghan governments' efforts to support ALP that might inform future endeavors to establish and sustain local security forces in difficult environments.

CHAPTER TWO

Cross-Cutting Obstacles

Despite their determination to help build the capacity of ALP and MOI, NSOCC-A and its Coalition partners have contended with some daunting obstacles in transferring full responsibility for the ALP program to the Afghan government. This chapter describes some of the cross-cutting barriers that have impeded MOI's ability to support ALP. They include the following:

- MOI's lack of control over local security officials
- GIRoA's lack of commitment to ALP
- Uneven Quick Reaction Force support to ALP
- Coalition reluctance to shift responsibility for ALP to MOI
- Uncertainty surrounding the future of ALP.

Without the removal of these basic obstacles, any progress made in developing effective processes for manning, training, equipping, and supplying the ALP force will remain tenuous.

Obstacle 1: MOI Does Not Control Local Security Officials

Although there is no consensus on what should be done about it, the lack of MOI control over those who command ALP is at the heart of many support problems that ALP guardians currently face. According to one Coalition adviser, the ALP director cannot provide necessary oversight over forces in the field because mainly district and provincial officials command those forces. While some local leaders use their authority to the benefit of ALP, others dole out the resources intended for ALP as they see fit, sometimes diverting them to other elements of ANP that they deem more in need or with which they have a tighter relationship—and sometimes simply refusing to release ALP supplies altogether.[1] ALP headquarters is officially powerless to do anything about such situations, their role being primarily that of an advocate rather than a resourcing and commanding organization.[2]

An ALP headquarters official we spoke to in 2013 said that the solution to the problem of recalcitrant or uncooperative local MOI officials was an independent chain of command connecting the national ALP leadership to ALP guardians in the field.[3] Although this idea once resonated within some in the SOF community, MOI and Coalition officials dismissed this proposal when it was made in 2012 and continued to oppose it when we spoke to them.[4] According to one NSOCC-A adviser, connecting ALP headquarters with local ALP officials may solve some of the logistics and support issues, but it would ultimately be a step back because ALP headquarters does not have the capability to oversee the "maneuver" aspects of ALP operations that are currently handled by provincial and district officials.[5] To reduce the potential for collusion and corruption at the provincial level, a Coalition adviser proposed in 2013 that the PCOP report directly to the MOI Deputy Minister of Security as opposed to the provincial governor. Yet this proposed reform raises issues of its own; some advisers questioned its legality and others claimed it was impractical because the Deputy Minister of Security's office would be unable to manage some 50 or more PCOPs.[6]

---

[1] NTM-A official, interview with the authors, February 17, 2013.

[2] NSOCC-A official, interview with the authors, February 25, 2015.

[3] MOI official, interview with the authors, February 19, 2013.

[4] NTM-A official, interview with the authors, February 17, 2013.

[5] NSOCC-A official, interview with the authors, February 26, 2015.

[6] NTM-A official, interview with the authors, February 22, 2013.

Little clarity exists regarding Afghan preferences for ALP headquarters. One NSOCC-A official reported in 2015 that he had not broached the subject of headquarters structure with senior Afghan officials, believing the Afghans would refuse to discuss the matter.[7] For its part, ALP headquarters has attempted to gain visibility of district-level operations and provide local guardians with the sense that ALP headquarters is a resource for coordination. ALP leadership has conducted staff assistance visits throughout the country to audit ALP accounts and inspect facilities and logistics, and has held security summits that bring together provincial and district officials, as well as ALP guardians.[8] However, ALP headquarters was not empowered to schedule or conduct staff assistance visits and summits on its own authority; such direction had to be issued in writing from superiors at MOI.

Obstacle 2: GIRoA May Lack Commitment to ALP

To some ALP officials and their Coalition advisers, the key issue has not been whether MOI has the capacity to support ALP, but whether it has the will to do so.[9] Structures (ministries, inspector generals, official codes of conduct) established by the Coalition mean little if the Afghans are not inclined to execute them,[10] and MOI's commitment to the ALP program has been lacking.[11] One NTM-A adviser in 2013 went so far as to say, "the minute [U.S.] SOF pulls out, the ALP will go away."[12] In 2015, another adviser noted that the Afghans had a decision to make on ALP in the next two years about whether to resource the

---

[7] NSOCC-A official, interview with the authors, February 26, 2015.

[8] According to one Coalition adviser, summits are more-significant events than staff assistance visits because representatives from each security pillar and other interested parties are brought together to discuss relevant issues (NSOCC-A official, interview with the authors, February 26, 2015).

[9] NSOCC-A officials, interviews with the authors, February 14 and 17, 2013.

[10] NTM-A official, interview with the authors, February 19, 2013.

[11] SOF team member, interview with the authors, February 17, 2013.

[12] NTM-A official, interview with the authors, February 22, 2013.

program though regular MOI channels. Although the United States has allocated funds to pay for ALP through 2018, policy changes in Washington or shifting U.S. priorities within Afghanistan could cause this funding stream to dry up.[13]

Ministerial attitudes toward ALP were difficult to gauge in an objective manner. When interviewed in 2013, some senior MOI officials did not feel they "owned" ALP in the way they did AUP. According to several NTM-A advisers, their Afghan counterparts in MOI "did not get the ALP"; they perceived it to be a foreign creation composed of poorly trained nonprofessionals.[14] This perception seems to have continued in 2015. One U.S. adviser argued that Afghan resistance to ALP comes from the fact that it is associated with the United States and not closely tied to the Afghan security structure. He further argued that those opposed to ALP favor a localized police force, but not one based on the SOF model of part-time village militia that are distinct from the full-time uniformed police.[15] The senior MOI officials to whom we spoke in 2013 indicated that they supported the continuation of the ALP program in the near term. However, they were vague about how long the program should continue and in what form. Although they said that the program was better managed than it had been, some expressed discomfort with ALP recruiting and operating practices. For example, one MOI general implied that it would be difficult to maintain a balance among tribes within ALP forces at the local level.[16]

MOI officials below the national level had mixed views regarding ALP. Reportedly, zone commanders expressed support for the ALP program at a 2013 conference attended by the ALP director, and ALP headquarters officials at the time pointed to a groundswell of demand for new ALP personnel in many provinces and districts because of the

[13] NSOCC-A official, interview with the authors, February 25, 2015.

[14] NTM-A officials, interviews with the authors, February 19 and 22, 2013.

[15] CSTC-A official, interview with the authors, February 25, 2015.

[16] MOI official, interview with the authors, February 18, 2013.

demonstrated security benefits of the program.[17] But one Coalition adviser attributed problems in distributing supplies to continued animosity between AUP and ALP at the local level.[18] The consequences of this lack of local support have been severe. Several Coalition advisers in 2015 described situations in which ALP guardians who felt unsupported reached agreements with insurgent forces, increasing the possibilities of insider attacks and breakdowns in police effectiveness.[19]

Obstacle 3: Quick Reaction Force Support to ALP Is Uneven

Perhaps more important than logistics support to ALP's success has been the operational support provided to lightly armed guardians by other elements of ANP and local units of ANA. When ALP forces face insurgent attacks that they cannot handle on their own, a Quick Reaction Force is supposed to be provided first by neighboring ALP forces, if available, then by AUP forces, and then by ANA forces if necessary.[20] In addition, ANA is supposed to ensure that ground lines of communication remain open in contentious districts and provinces.[21]

In 2013, ALP had demonstrated the capacity to call in AUP help when needed and had conducted successful joint operations with other ANP organizations and ANA forces.[22] But the effectiveness of Quick Reaction Forces varied by location.[23] For example, a Special Operations Task Force East official contended that AUP and Afghan Border Police in his region seldom responded to ALP requests for Quick Reaction

---

[17] MOI official, interview with the authors, February 19, 2013.

[18] Regional Command East official, interview with the authors, February 13, 2013.

[19] NSOCC-A official, interview with the authors, February 25, 2015.

[20] NSOCC-A official, interview with the authors, February 14, 2013.

[21] NTM-A officials, interviews with the authors, February 22 and 23, 2013.

[22] SOF team members, interviews with the authors, February 15 and 23, 2013.

[23] NSOCC-A official, interview with the authors, February 16, 2013.

Forces.[24] In addition, a senior MOI leader castigated ANA for its inadequate Quick Reaction Force assistance to ALP, claiming that ANA units were too tied to Coalition forces and would not operate without them.[25] Despite a history of poor ministerial cooperation between MOI and the Ministry of Defense,[26] the MOI deputy minister, the chief of ALP, and the Ministry of Defense operations director had met to discuss challenges of getting Ministry of Defense support for ALP.[27] However, it was unclear whether the ministry had committed itself to better operational support to ALP, much less whether it would follow through on such commitments.

Obstacle 4: Coalition Reluctance to Shift Responsibility to MOI Deterred the Transition

U.S. and other Coalition SOF support has been key to the formation, growth, and effectiveness of ALP. Yet, ALP's continued reliance on the Coalition during the period of rapid force reductions and handover of most security responsibilities to the Afghan National Security Forces concerned some officials we spoke to in 2013 and 2015. According to one interviewee, ALP guardians were more loyal to SOF than to GIRoA.[28] The fact that salaries and stipends for ALP were paid for by U.S. funds—hence segregated from other funding streams in MOI—was not lost on MOI officials and administrators. One Coalition official even wondered in 2015 whether continued championing and oversight of ALP by Coalition SOF had forestalled the program's integration with ANP, which was CSTC-A's responsibility to advise and assist. He asked rhetorically, "Why is ALP working with SOF? Right now, [CSTC-A has] no relationship with ALP," even though CSTC-A

---

[24] NSOCC-A official, interview with the authors, February 16, 2013.

[25] MOI official, interview with the authors, February 21, 2013.

[26] NTM-A official, interview with the authors, February 17, 2013.

[27] MOI official, interview with the authors, February 19, 2013.

[28] NSOCC-A official, interview with the authors, February 23, 2013.

worked well with other specialized police forces that were also advised by Coalition SOF.[29] Despite these reservations, it seems unlikely that any group other than Coalition SOF will be given primary responsibility for mentoring the ALP program. Furthermore, ALP officials and Coalition advisers believed in 2013 that persistent Coalition SOF engagement with MOI, and additional training, would be critical to ALP's future, and this was even more true in 2015 following the drawdown of Coalition combat forces.[30]

In 2015, Coalition interviewees voiced the opinion that many of the issues faced by ALP could be traced to funding. As one official explained, the GIRoA depends on donations for its security functions, and in ALP's case, the funds come entirely from the U.S. government.[31] This is in contrast to the rest of ANP, which was funded at the time of our interviews by the United Nations–administered Law and Order Trust Fund for Afghanistan. This had several effects. First, as previously discussed, there was a sense among some Afghan officials that the United States would resolve any ALP issue that arose. Second, local Afghan leaders at times withheld required services or food from guardians on the belief that they would be supplied by the United States.[32] One NSOCC-A adviser stated that the Law and Order Trust Fund for Afghanistan might be discontinued and, in any event, may not be the appropriate funding mechanism for ALP. However, he remained hopeful that international donors will be willing in the future to take on some of the ALP funding responsibilities, given the overall beneficial effect of this force on local security and its relatively low cost compared with other components of ANP and ANA.[33]

---

[29] NSOCC-A official, interview with the authors, February 25, 2015. He mentioned, in particular, CTSC-A's good working relationship with the General Command of Police Special Units, a command section of MOI These police units include Provincial Special Units and national-level special police units; special operators in NSOCC-A advise the units.

[30] NTM-A officials, interviews with the authors, February 18 and 19, 2013.

[31] NSOCC-A official, interview with the authors, February 25, 2015.

[32] CSTC-A official, interview with the authors, February 25, 2015.

[33] NSOCC-A official, interview with the authors, February 25, 2015.

Obstacle 5: The Future of ALP Is Unclear

MOI officials are at a critical decision point for the future of ALP. With personnel levels stable around 30,000 and the Coalition pledging funding through 2018, though decreasing its overall presence, the Afghan government will need to improve its support capabilities, merge ALP with AUP, or dissolve the program. In 2013, MOI leaders stated that the status of ALP as an independent entity was entirely dependent on the security situation in Afghanistan.[34] If the situation stayed the same or worsened, ALP would likely continue in its current form as a semi-autonomous entity within AUP. If the situation got better, then ALP could be disbanded or completely absorbed into ANA, ANP, or other agencies.[35] The position of the United States in 2015 was that a need for the type of local defense provided by ALP would persist beyond 2018, but this did not necessarily mean that ALP should continue as a separate subpillar of AUP.[36] In 2013, Coalition officials indicated that disbanding ALP could be a dangerous course of action in the absence of a plan for disarmament, demobilization, and reintegration.[37]

Assuming that ALP will continue as the Resolute Support mission evolves and Coalition forces continue to draw down, will they also continue to conform structurally and functionally to the vision of Coalition SOF? As demonstrated in the 2012 negotiations over future tashkil levels, the Afghans have their own views of where ALP should be located and in what numbers, which do not always align with Coalition strategic priorities. It was apparent then that some districts with authorized ALP lack the population base or the popular support to host the number of forces they were originally allocated, whereas other districts were clamoring for more guardians than their current tashkils permit.[38] At times, this resulted in "harvesting" tashkil in one district to meet the needs of another. At the macro level, some Afghans stated

---

[34] MOI officials, interviews with the authors, February 18 and 21, 2013.

[35] MOI officials, interviews with the authors, February 19 and 21, 2013.

[36] NSOCC-A official, interview with the authors, February 25, 2015.

[37] NTM-A official, interview with the authors, February 21, 2013.

[38] NSOCC-A official, interview with the authors, February 15, 2013.

that the 2013 goal of 45,000 ALP personnel was woefully inadequate, with one regional official at the time recommending that the force level be raised to 70,000.[39] Although unrealistic even then, such an outsized desire when compared with others' skepticism regarding the need for any ALP at all in the future indicates a considerable variation in official Afghan perspectives on ALP that continue to this day. Another issue is the geographic disposition of ALP sites. Since the inception of the program, some local Afghan power brokers have used the ALP program for political patronage (mostly in the northern part of the country, but also in the eastern province of Nangahar). One U.S. official interviewed in 2015 referred to the danger of patronage with such a group by describing how a comparable Russian effort produced armed militias for warlords during the 1980s.[40]

Afghan and Coalition views have also differed regarding the appropriate role for ALP in facilitating local security, as well as the extent of ALP integration into AUP. In accordance with SOF doctrine, ALP forces were designed to be part-time, village-based security guards with limited armament and mobility and no law enforcement powers, except detention authority.[41] Furthermore, they were tethered to the government through the DCOPs. Progressively, however, being an ALP guardian has turned into a full-time occupation for many, and some ALP units have operated outside their villages, mostly at checkpoints, in cooperation with AUP.[42] To one MOI official interviewed in 2013, this was a positive development, because it reduced the security burden on the rest of ANP and allowed a more flexible employment of resources.[43] But other Afghans and their Coalition advisers were leery of ALP forces undertaking more AUP-type work away from their vil-

[39] MOI official, interview with the authors, February 14, 2013.

[40] CSTC-A official, interview with the authors, February 25, 2015.

[41] MOI official, interview with the authors, February 21, 2013.

[42] MOI Official, interview with the authors, February 21, 2013.

[43] MOI official, interview with the authors, February 18, 2013.

lages, contending that maintaining ALP forces on-site, rather than deploying them, minimizes ALP casualties.[44]

In 2013, a commonly held view among MOI officials and some Coalition advisers was that ALP should be fully integrated into the regular police force in the not-too-distant future. From this perspective, ALP would become the entry point to AUP and "form the foundation of community policing."[45] In a 2013 ANP vision statement, MOI mentioned a plan to further integrate ALP into ANP, asserting, "Within ten years, as security conditions improve, the ALP will be transformed." The implication was that "after receiving the required education and training," ALP guardians would be virtually indistinguishable from members of ANP.[46] To proponents, such integration or assimilation could help counter a trend toward increased autonomy of local security forces following the Coalition's exit from rural areas of the country. But this view of ALP's future was not a universal one. In 2013, some Coalition advisers contended that the wholesale conversion of ALP to AUP would be cost-prohibitive, while others found existing procedures for ALP forces to enter ANP and ANA entirely adequate.[47] Despite the diversity of views regarding ALP's future, the message delivered to MOI leadership by the Coalition in early 2015 was to get through the "fighting season," develop a conditions-based drawdown plan, assess results with hard metrics, and end the ALP program in areas where appropriate, such as where there was limited insurgent activity or where ALP had been consistently compromised.[48]

---

[44] Regional Command East official, interview with the authors, February 13, 2013; MOI official, interview with the authors, February 18, 2013; and NTM-A official, interview with the authors, February 21, 2013.

[45] NTM-A officials, interviews with the authors, February 19 and 21, 2013.

[46] Islamic Republic of Afghanistan, 2013b.

[47] MOI officials, interviews with the authors, February 19 and 25, 2013; and NSOCC-A officials, interviews with the authors, February 19 and 25, 2013.

[48] NSOCC-A official, interview with the authors, February 26, 2015.

Conclusion

The Coalition has faced major obstacles in its effort to help MOI take ownership and control of the ALP support enterprise. Some of these obstacles extend across the MOI functions that sustain the ALP program. Such obstacles include the lack of MOI and ALP headquarters authority over PCOPs, the uncertain commitment of senior MOI leaders to the ALP program, the difficulty of assessing conditions in ALP districts and accounting for supplies and personnel, the reluctance of Coalition forces until recently to permit MOI to assume responsibility for the full range of ALP support functions, and the uncertainty surrounding the role of ALP within MOI's future policing strategy.

Aside from these cross-cutting issues, ALP program stakeholders have also wrestled with support-related problems in logistics, personnel management, and training. These issues are addressed in the next three chapters.

CHAPTER THREE

Logistics

With Coalition help, MOI has made significant strides to improve its logistics practices and results.[1] Yet much more needs to be done if the Afghans are to acquire a full and independent capacity to plan for, manage, requisition, track, store, transport, distribute, and maintain necessary quantities of ALP equipment and supplies; to build and maintain sufficient numbers of ALP checkpoint facilities; and to support the medical and communications requirements of ALP personnel.

The MOI General Directorate of Logistics has responsibility for providing all equipment and materials necessary for ALP and its chain of command.[2] Historically, Coalition forces provided critical logistics support to MOI in order to ensure that ALP was supported in the field. The results of this combined effort have been uneven, with ALP guardians in some provinces and districts receiving most of their authorized equipment and supplies and others faring considerably worse. As of

[1] This chapter uses a rather expansive definition of *logistics*. Generally, it conforms to the NATO definition, which describes logistics as the "the science of planning and carrying out the movement and maintenance of forces." This includes the "design and development, acquisition, storage, transport, distribution, maintenance, evacuation and disposal of materiel; acquisition or construction, maintenance, operation and disposition of facilities; transport of personnel; acquisition or furnishing of services; and medical and health service support." See NATO, *NATO Logistics Handbook*, Brussels, Belgium, October 1997. We have added communications to this list of functions in line with the United Nations' definition of logistics.

[2] This includes combat materials, technical supplies, and organizational clothing and individual equipment in the ALP tashkil (Deputy Commander of Special Operations Forces, NTM-A, 2012c).

early 2015, ALP was entirely reliant on the MOI and AUP logistics system, with the exception of communications gear and organizational clothing and individual equipment. Once the final orders of these classes have been filled, they will also shift to resupply via the AUP system.[3] Complicating the resupply process, ALP headquarters has no logistics representatives in the field, including at the regional logistics centers (RLCs), and there are no plans to create any such billets.[4]

Push or Pull Logistics System

The debate about whether the Coalition should emphasize a so-called *push strategy* (defined as supply based on forecasted demand) or a *pull strategy* (defined as supply based on actual or consumer demand) for MOI's supply chain management has been ongoing since ALP's creation. After spending the formative years of their careers using the Soviet system, many high-ranking MOI logistics officers understand the push method fairly well. However, this process has proven unable to provide timely and accurate reports of initial equipment and resupply needs. For example, in late November 2012, NTM-A identified that equipment for about 2,000 more guardians than authorizations had been pushed to ALP units in the field.[5] Attempting to mitigate this imbalance, NSOCC-A decided to execute pushes based on the number of guardians on hand rather than the number authorized in the tashkil.[6] However, the new push policy was unable to counteract problems with "overburdened and unreliable convoys" and "loss, theft, or obstruction in transit."[7] In addition, MOI favored pushes based on authorizations rather than actual fill levels. As one NTM-A official put it, the MOI National Logistics Center took a "let's get equipment

---

[3] NSOCC-A official, email to the authors, March 9, 2015.

[4] NSOCC-A official, interview with the authors, February 25, 2015.

[5] NSOCC-A, "Information Paper, 20 Feb 2013," CJ4, Kabul, Afghanistan, 2013b.

[6] NSOCC-A, 2013b.

[7] NSOCC-A, 2013b.

out to these guys [the Afghan National Security Forces] as quickly as possible" approach.[8] This intense desire to move equipment at the national level, combined with a lack of faith in the capability of the Afghan supply system and a desire to use ALP supplies for other purposes, reportedly led to extensive hoarding and diversion of equipment at provincial and district levels in 2013.[9] The reduction in the number of ALP personnel authorized in 2014, from 45,000 to 30,000, made initial equipment targets more attainable and, presumably, the logistics process somewhat less prone to leakage and corruption. As of March 2015, NSOCC-A indicated that ALP vehicles, communication gear, and individual equipment were being brought to 100-percent fill.[10]

In 2015, MOI formally shifted to a pull system—that is, a system in which supplies are moved only after the submission of the required documentation from field units—to meet ALP resupply and replacement needs.[11] Although the Coalition has taken steps to facilitate this transition—such as providing logistics training and advice to ALP headquarters staff members and encouraging them to train and evaluate provincial- and district-level officials responsible for ALP logistics—establishing a fully functioning pull system in Afghanistan faces daunting obstacles, some of which are discussed below.

---

[8] NTM-A official, interview with the authors, February 20, 2013.

[9] Although a less serious problem than hoarding, pilfering and diversion of equipment meant for ALP is also a regular occurrence, according to Coalition officials. In some cases, PCOPs and DCOPs are unaware of which shipments belong to ALP and other elements of ANP. In other cases, they consider what they receive as theirs to keep or further allocate as they see fit. Coalition advisers indicated that such problems are not uniform throughout Afghanistan. In some areas, ALP guardians experience few problems in getting their authorized equipment. Indeed, they sometimes are issued weapons more offensive and advanced than those they were intended to receive. (NSOCC-A officials, interviews with the authors, February 13, 14, 16, 20, 24, and 25, 2013)

[10] NSOCC-A official, email to the authors, March 9, 2015.

[11] NSOCC-A official, interview with the authors, February 25, 2015.

Initial Equipment Issue

In the latter half of 2012, NSOCC-A and NTM-A undertook an initiative to eliminate the bottlenecks in MOI's logistics process that were holding up deliveries of initial equipment sets to ALP trainees in newly validated districts—and simultaneously, to improve the capacity of MOI to manage this process itself. As indicated above, the Coalition had to modify its push approach to fit Afghan circumstances, yet the goal of providing ALP guardians with their initial equipment was ultimately achieved, in large part by furnishing new recruits with their kits at RTCs.

Coalition Support

In early 2013, Coalition assistance proved necessary in light of difficulties that MOI faced in providing ALP guardians with their authorized equipment, which adversely affected ALP's "operational deployment, readiness, personnel safety, and ability to function in adverse weather."[12] As the ALP program was getting off the ground, Coalition forces frequently provided direct logistics support. In order to ensure that ALP personnel in the field received critical supplies, Coalition forces transported designated "trusted agents" to RLCs—or, depending on location, to the national logistics center in Kabul—to pick up and distribute ALP equipment. These agents verified and signed for equipment and supplies as they were loaded and sealed in containers for shipment,[13] circumventing MOI officials at the provincial and district levels. While this approach was "not ideal for building a sustainable ANP logistical system," it helped to limit pilfering and ensure timely and secure equipment delivery.[14]

The Coalition's hands-on approach to the ALP support problem had its detractors and proponents among MOI officials and advisers. Some senior MOI interviewees took issue with Coalition forces giving

---

[12] U.S. Department of Defense Inspector General, 2012.

[13] NSOCC-A, *Afghan Local Police Pay and Logistics Guide*, Kabul, Afghanistan, May 2013i.

[14] U.S. Department of Defense Inspector General, 2012.

weapons directly to ALP forces rather than through MOI channels.[15] Other MOI advisers lamented the Coalition's strategy of pushing ALP equipment into the field as soon as it was procured. They recommended that the U.S. government not send over equipment until Afghans were ready to use it; otherwise, it would likely sit in the desert, where it would degrade or be stolen.[16]

By 2015, the ALP logistics process was largely turned over to MOI, where it was managed as part of the AUP supply system. The U.S. government still played a role in contracting for certain items, such as uniforms, for which it faced some criticism because of slow fill times by U.S. manufacturers.[17] However, Coalition advisers reported that equipment distribution at the RTCs was generally satisfactory.[18]

Logistics Shuras

Throughout 2012, NSOCC-A, in cooperation with NTM-A and ALP headquarters, convened national and regional logistics *shuras* (or conferences) to help resolve logistics problems related to initial equipment issue and resupply. These shuras focused on helping ANP become more self-sufficient by emphasizing the use of existing supply chains and maintenance channels and by enhancing asset management.[19] The shuras were able to resolve immediate equipment backlogs by opening lines of communication between key logistics and operational personnel throughout the AUP and ALP command structure. Partly as a result, regional advisers reported that most newly validated ALP sites were receiving their authorized "shoot, move, communications" packages in a timely manner.[20] But Coalition advisers had not only initiated the logistics shuras, they had actively supported them by transporting

---

[15] MOI officials, interviews with the authors, February 18 and 19, 2013.

[16] NTM-A officials, interviews with the authors, February 18 and 20, 2013.

[17] NSOCC-A official, interview with the authors, February 25, 2015.

[18] NSOCC-A official, interview with the authors, February 25, 2015.

[19] Ashley Curtis, "Logistics Shura Encourages ANP Self-Sufficiency," Kandahar Airfield, Afghanistan, 117th Mobile Public Affairs Detachment, June 10, 2012.

[20] NSOCC-A official, interview with the authors, February 16, 2013.

ALP headquarters officials to scheduled events throughout the country. Although the RAND team witnessed a well-attended and mostly Afghan-led regional logistics conference in Gardez, it was unclear in 2013 whether MOI had the willingness or capability to convene such conferences on a regular basis without Coalition involvement.

"ALP in a Box"

In October 2012, NSOCC-A and NTM-A established the "ALP in a box" concept. According to the plan, preconfigured clothing and equipment shipments would be packaged and sealed at the Materiel Management Center in Kabul and then shipped to RLCs, where they would remain until a trusted agent from the receiving ALP unit showed up to sign for that unit's boxes. These boxes included organizational clothing and individual equipment. Depending on the unit's needs, they also contained weapons, ammunition, and medical supplies. The objective was to prevent pilfering and diversion of supplies as materials passed through provincial and district headquarters en route to ALP sites.

Designed as a temporary measure, the concept seemed to work well when tested under favorable logistical conditions in northern Afghanistan, but failed when put to the test in an environment where the physical and human infrastructure was less robust. According to one Coalition official, getting all the equipment in the same place at the same time was difficult without reliable transportation and trained and trustworthy logisticians.[21] For instance, a box might show up with a single item, such as gloves, rather than a full complement of gear. While the concept made sense in theory, Coalition logisticians determined that it "was slowing things down" and not "Afghan sustainable."[22] This was an unfair assessment to some who thought that the concept had not been fully tested or given the opportunity to succeed under a range of conditions. Though never officially canceled, ALP in a box was shelved

---

[21] NTM-A official, interview with the authors, February 20, 2013.

[22] NSOCC-A official, interview with the authors, February 14, 2013.

in early 2013 in favor of a return to standard logistics practices.[23] The long-term solution to the initial equipping problem, according to one Coalition official, was to provide ALP guardians with their gear while they were being trained at RTCs.[24]

Resupply Process

By 2015, the ALP initial equipping process seemed to be performing effectively under Afghan management. But the resupply situation was more problematic. In some cases, the Afghans' inability to master or follow the steps of MOI's requisition process has led to problems in ordering the equipment and supplies authorized for ALP. In other cases, it has led to supplies being diverted or hoarded prior to reaching ALP personnel in the villages; this diverting and hoarding primarily happens after the supplies leave the RLCs. One official in 2013 expressed concern that ALP would not be able to obtain adequate supplies of fuel and ammunition as the Coalition presence waned.[25]

ALP units have often faced challenges in the "pull" logistics system. In many cases, advisers noted in 2013, they simply "don't know the system," in terms of both what they can ask for and how to ask for it.[26] For example, guardians might not understand how to fill out the consumption reports that are required to accompany their requests for fuel and ammunition. In such cases, district officials might not forward ALP request forms to higher headquarters, making it difficult for regional logistics coordinators to ascertain lower-level demand.[27] In general, guardians go to the district commander for supplies and only turn to ALP headquarters if there is a problem. According to one

---

[23] NTM-A official, interview with the authors, February 25, 2013; and U.S. Department of Defense Inspector General, 2012.

[24] NSOCC-A official, interview with the authors, February 15, 2013.

[25] NSOCC-A official, interview with the authors, February 16, 2013.

[26] NSOCC-A official, interview with the authors, February 17, 2013.

[27] MOI official, interview with the authors, February 14, 2013.

Coalition adviser we interviewed in 2015, sometimes the problem gets solved, sometimes not.[28]

Sometimes, the problem is not a matter of ALP ignorance of the requisition system, but rather district or provincial leaders' reluctance to provide ALP forces with substantial supplies, stemming from those leaders either having other uses in mind for the supplies or believing the guardians will misuse them unless allotted sparingly.[29] One U.S. official stated in 2015 that ALP had been done a great disservice by not being institutionalized within MOI much faster, because provincial and district officials now see them as "resource fodder," using them to justify requests for equipment that ALP will never see.[30] Finally, several Coalition officials indicated in 2013 that MOI's Western-oriented pull process for requesting supplies went against the Afghan grain, and that a Soviet-style push system is better suited to a security force with limited administrative capacity.[31] Indeed, some predicted that the Afghans would revert to Soviet logistics practices once Coalition forces left the country.[32]

Equipment and Resupply Results

In 2013, Coalition officials provided mixed reviews of MOI's logistics capabilities and potential. According to one MOI adviser, the ALP logistics process was stuck in a top-down Soviet-style system, which made it difficult for Afghans to push and pull equipment and supplies in a timely manner.[33] Other Coalition officials were more positive. As they saw it, the Afghan logistics process was "full of corruption" and

---

[28] NSOCC-A official, interview with the authors, February 25, 2015.

[29] NSOCC-A official, interview with the authors, February 14, 2013.

[30] CSTC-A official, interview with the authors, February 25, 2015.

[31] NSOCC-A officials, interviews with the authors, February 17 and 18, 2013; and NTM-A officials, interviews with the authors, February 17 and 18, 2013.

[32] SOF team member, interview with the authors, February 2013.

[33] NTM-A official, interview with the authors, February 22, 2013.

"not 100-percent effective." But it worked well enough at the time, and should perform adequately in the future without direct Coalition oversight and assistance.[34] Many officials indicated that the extent of MOI logistics support provided to ALP depended on the local situation. For example, while the distribution of supplies was good in Herat, where the security situation was relatively stable, it was a "mess" in Helmand, where security conditions made transporting supplies more challenging.[35] Coalition and Afghan interviewees said that the logistics chain from MOI's national warehouses in Kabul to the zone-level RLCs functions relatively well;[36] however, the chain tended to break somewhere between the RLCs and ALP forces in the villages, with some pointing to the provincial headquarters and some to the district headquarters as the point of maximum disruption.[37] According to these officials, the organization of MOI at the provincial and district levels made the success or failure in provisioning ALP forces largely a matter of the "personalities" of the particular PCOPs and DCOPs.[38]

To some interviewees, the existence of troublesome links in the logistics chain indicated that ALP should be given the opportunity to "cut out the middle man" by going directly to the RLCs for needed supplies and equipment.[39] To some degree, this reform has been implemented. In 2013, ALP guardians began to be equipped at provincial and regional training centers as they rotated through these facilities and Coalition advisers.[40] When operating in the field, however, guard-

[34] NSOCC-A officials, interviews with the authors, February 13 and 24, 2013.

[35] NSOCC-A official, interview with the authors, February 17, 2013.

[36] SOF team member, interview with the authors, February 2013; and NSOCC-A official, interview with the authors, February 17, 2013.

[37] SOF team member, interview with the authors, February 2013; NSOCC-A officials, interviews with the authors, February 13, 15, 16, and 20, 2013; and NTM-A officials, interviews with the authors, February 20, 2013.

[38] SOF team member, interview with the authors, February 2013; NSOCC-A officials, interviews with the authors, February 14, 16, 17, 24, and 25, 2013; and NTM-A officials, interviews with the authors, February 17 and 25, 2013.

[39] NSOCC-A official, interview with the authors, February 16, 2013.

[40] NSOCC-A official, interview with the authors, February 25, 2015.

ians are required to submit their supply requests to their district commander, who works through the district and provincial leadership to fulfill them. Thus, some of the problems identified in 2013 continued to exist in 2015, with some PCOPs and DCOPs withholding supplies, skimming supplies, or using ALP requests to obtain supplies that are then given to other units.[41] At times, this has prompted ALP commanders to call either ALP headquarters or, in some cases, the ALP SOAG for assistance. However, both groups are limited in their ability to influence what happens within a PCOP's or DCOP's jurisdiction.[42] Coalition advisers noted, however, that they had not observed degradation in ALP performance from a lack of equipment in the field. A more pervasive problem was moving ALP guardians away from their villages and their existing lines of supply—the system is designed for the guardians to eat and sleep at home—without providing an alternative means of supporting them.[43]

Accountability

The actual amount of hoarding, diversion, and outright theft of ALP equipment is difficult to determine, especially now that the Coalition has very little visibility on ALP units in the field. While one Coalition official contended in early 2013 that the MOI requisition system then in place was "good enough" for accountability purposes,[44] another in 2015 believed that it was inadequate.[45] During an interview in 2013, we learned that the MOI's main munitions warehouse in Kabul had not conducted a complete inventory of its holdings.[46] That said, a Coalition adviser we spoke to indicated that the national logistics center was doing a reasonably good job of tracking the ALP equipment pushed

[41] NSOCC-A official, interview with the authors, February 25, 2015; and CSTC-A official, interview with the authors, February 25, 2015.

[42] NSOCC-A official, interview with the authors, February 25, 2015.

[43] NSOCC-A official, interview with the authors, February 25, 2015.

[44] NSOCC-A official, interview with the authors, February 16, 2013.

[45] NTM-A official, interview with the authors, February 20, 2013.

[46] NTM-A official, interview with the authors, February 25, 2013.

out to the RLCs.[47] But MOI was ineffective in accounting for supplies once they left the RLC,[48] which one Coalition official attributed to MOI's lack of digital tracking systems.[49]

Prior to 2014, the Combined Joint Special Operations Task Force–Afghanistan (CJSOTF-A) and its subordinate special operations task forces tried to compensate for this shortcoming by creating a shadow tracking mechanism. Yet this mechanism focused on the district and ignored the key provincial waypoint between the RLC and the district. In addition, the accuracy and reliability of CJSOTF-A's logistics information was open to question, given that it relied on Afghan sources of information and was only partially verified.[50] Furthermore, as some Coalition officials recognized, the ultimate goal ought to be an Afghan-sustainable tracking and accountability mechanism for logistics. And to realize that goal, one adviser suggested, the Coalition should encourage MOI to build on its existing paper-based supply system rather than attempt to impose a computer-based system that the Afghans are unable to use.[51]

Bottom Line

According to Coalition advisers in 2013, the bottom line was that some ALP units were meeting their requirements with and without MOI support,[52] but others faced serious supply and equipment shortages.[53] According to one Coalition official, certain ALP units would not be able to sustain operations in Regional Command East without additional ammunition.[54] At the regional logistics conference attended by

---

[47] NTM-A official, interview with the authors, February 18, 2013.

[48] NSOCC-A official, interview with the author, February 13, 2013.

[49] Regional Command East official, interview with the authors, February 13, 2013.

[50] SOF team member, interview with the authors, February 2013; and NSOCC-A official, interview with the authors, February 13, 2103.

[51] NTM-A official, interview with the authors, February 25, 3013.

[52] NSOCC-A officials, interviews with the authors, February 13 and 23, 2013.

[53] NSOCC-A official, interview with the authors, February 16, 2013.

[54] Regional Command East official, interview with the authors, February 13, 2013.

the RAND team in Gardez in February 2013, Afghan officials from Paktika, Khost, and Wardak provinces complained about a lack of spare parts.[55] In 2013, a senior MOI official in Kabul mentioned that ALP guardians were in need of uniforms, food, and firewood, particularly for those stationed at some distance from their villages.[56] Probably of greatest concern to MOI leaders was ALP's lack of appropriate weapons. Too many ALP guardians in their view were receiving inferior Czech-made AK-47 rifles, even though they were supposed to have the Russian version, a concern that, by 2015, had not been addressed.[57] This situation may be relieved as MOI fields Western-made weapons, thus making higher-quality AK-47s available for issuing to guardians.[58] Concerns were also raised in 2013 about whether ALP guardians were appropriately armed. Believing ALP to be outgunned by the insurgents, one MOI official decried the guardians' lack of heavy weapons.[59] But it was unlikely that ALP would see any heavy weapons in the future, at least through official channels. According to one Coalition official, the effort to brand ALP as a force worthy of support from the international community would suffer if ALP were seen as having an offensive capability.[60]

A RAND study team analysis of ALP logistics data for January 2012 through January 2013, supplied by NSOCC-A, showed mixed results for the 21 districts whose ALP programs had been under Afghan control for at least three months, albeit with some degree of SOF oversight—that is, for the mature tactical overwatch districts. On average, ALP units in these districts had almost all of their authorized complement of light weapons (AK-47s), but the median level of machine guns, ammunition, and trucks was less than 50 percent. The

[55] MOI official, statements at a conference attended by the authors, February 13, 2013.

[56] MOI official, interview with the authors, February 21, 2013.

[57] MOI officials, interviews with the authors, February 25, 2013; and NSOCC-A officials, interviews with the authors, February 19 and 25, 2013.

[58] CSTC-A official, interview with the authors, February 25, 2015.

[59] MOI official, interview with the authors, February 14, 2013.

[60] NSOCC-A official, interview with the authors, February 26, 2015.

situation was even worse in the 38 non–tactical overwatch "priority districts" that we examined,[61] where the median level of machine guns and ammunition was approximately 20 percent.[62] After controlling for factors related to time, topography, demographics, and stability, we still found statistically significant differences between mature tactical overwatch and non–tactical overwatch districts. On average, the percentage of authorized machine guns was 32 percent higher and the percentage of authorized ammunition was 29 percent higher in the former than in the latter. Although differences between the two categories of districts also existed for trucks and light weapons, they were not statistically significant. Presumably, the push by Coalition SOF to supply ALP prior to fully handing over the support mission to MOI explains much of the equipment gap between tactical overwatch and non–tactical overwatch districts in 2012 and 2013, but it does not explain why so many mature tactical overwatch districts faced substantial equipment shortages.

In his October 2015 report, the U.S. Department of Defense Special Inspector General for Afghanistan Reconstruction found MOI logistics support to the ALP "unreliable and insufficient." In particular, ALP commanders interviewed by his staff said that the fuel they ordered, "usually arrived late and at an amount less than requested." The report also referenced an August 2014 inventory review and field inspection by ALP headquarters and coalition personnel that uncovered "significant shortages" of weapons, trucks, and motorcycles in multiple districts in Helmand province.[63]

---

[61] These priority districts were designated such by CJSOTF-A.

[62] NSOCC-A, ALP data provided to RAND, January 2012–January 2013. RAND's analysis of ALP logistics was constrained by the quality of data on ALP units at the district level. These data were supposed to be collected on a weekly basis in each district. However, data for all variables were sometimes unavailable for particular weeks. In addition, many consecutive weeks contained the same values. To deal with the availability and variability issues, we aggregated data for each district to the monthly level. Then we regularized the data using measurement standards established by the Coalition and MOI.

[63] Special Inspector General for Afghanistan Reconstruction, *Afghan Local Police: A Critical Rural Security Initiative Lacks Adequate Logistics Support, Oversight, and Direction*, Arlington, Va., SIGAR 16-3 Audit Report, October 2015b.

Other Logistics Functions

Transportation

In the words of one veteran ALP administrator in 2013, the ability to supply equipment and provide trainers for ALP was largely "a transportation and security problem."[64] By circumstance and design, ALP formation has often occurred in remote regions of Afghanistan that constitute favorable human and physical terrain for insurgents and a harsh operating environment for most Afghan National Security Forces. Without the technological and logistical advantages offered by the Coalition, the Afghan government faces two difficult tasks with respect to ALP: transporting supplies down to the village level and transporting guardians to and from RTCs.[65] Key to the resupply of ALP units in the field is the involvement of the aforementioned trusted agent or logistics officer. Before equipment can be offloaded from the ANP transportation brigade or contracted carrier transports, the Afghan representative has to call the administrative customer care number with tracking numbers to verify the integrity of the shipment.[66]

MOI's lack of airlift exacerbates the transportation problem, and it has had to develop ways to supply ALP units throughout the country over land. Despite hope that the Afghan air force will be able to perform airdrops in remote areas when it is more fully established, ground transportation will remain the only means of resupply for the foreseeable future.[67] The frequent fuel shortage faced by MOI and Ministry of Defense units is an additional concern.[68] Given sporadic Afghan government funding and Coalition reluctance to supply fuel, operational units have been prioritized over transport units, leading to shortages within ALP and AUP. In 2013, a senior Coalition official pointed to the standup of two of three planned transportation brigades as evi-

[64] MOI official, interview with the authors, February 19, 2013.

[65] NSOCC-A officials, interviews with the authors, February 16 and 17, 2013; and NTM-A officials, interviews with the authors, February 17, 2013.

[66] NSOCC-A, 2013i.

[67] NSOCC-A official, interview with the authors, February 14, 2013.

[68] NSOCC-A official, interview with the authors, February 25, 2015.

dence that MOI's ground transportation capability was growing.[69] Yet these brigades are intended to travel routes between the national warehouses and the RLCs.[70] They will not resolve the safety issues involved in transporting supplies from the RLCs to the villages, for which the only solution at present is for ALP and other ANP units to "brave" the roads in order to collect their provisions. Of course, many already do this and no doubt will continue to do so.

Equipment Storage

MOI's ability to store equipment is limited at the provincial level and below. In some locations, this results from improper utilization of existing capacity and the Afghans' inclination to hoard.[71] In other cases, capacity has been kept small to discourage corruption.[72] According to one Coalition adviser in 2013, the lack of local storage capacity put ALP at significant risk because forces could not store ammunition for contingencies.[73] Others doubted that storage limitations had much operational impact,[74] or indicated that salience of the storage issue depended on the location of the ALP unit. As one official put it, "The closer districts are to RLCs, the less storage matters."[75]

Equipment Maintenance

Coalition officials interviewed in 2013 stated that MOI lacked personnel with the technical and literacy skills required to maintain the vehicles belonging to various ANP elements, including the ALP program. Consequently, the Coalition outsourced MOI vehicle maintenance to a contractor, Automotive Management Services. Yet MOI logisticians and leaders at all levels questioned this contractor's capac-

---

[69] NTM-A official, interview with the authors, February 22, 2013.

[70] NTM-A official, interview with the authors, February 20, 2013.

[71] Regional Command East official, interview with the authors, February 13, 2013.

[72] NTM-A official, interview with the authors, February 22, 2013.

[73] NTM-A official, interview with the authors, February 17, 2013.

[74] NTM-A official, interview with the authors, February 22, 2013.

[75] NSOCC-A official, interview with the authors, February 24, 2013.

ity to deliver timely and reliable service, particularly in rural areas, and demanded that the contractor send out more mobile repair teams.[76] Although it was the district's responsibility to pick up or coordinate the transportation of its equipment,[77] many districts lacked the capability to deliver and retrieve equipment to and from distant maintenance locations. At the logistics conference that the RAND study team attended in Gardez in 2013, one provincial logistics chief stated that he had to send vehicles to Kabul to be fixed properly, and another indicated that more than a third of his trucks were not mission capable.[78] Backed by Coalition officials, Automotive Management Services representatives countered that ANP personnel did not perform preventive vehicle maintenance, drove their vehicles recklessly, and expected more from the contractor than it was required to deliver (e.g., a new tire rather than a repaired tire).[79] Still, the contractor acknowledged that the security situation constrains its ability to provide maintenance services in the hinterlands.[80] One Coalition adviser stated that ALP units themselves needed mechanics, with access to spare parts, so that they could perform basic repairs without having to rely on vehicle maintenance contractors at the provincial headquarters or RLC.[81]

Although one MOI official contended in 2013 that the maintenance function should be incorporated into his ministry's organizational structure,[82] the maintenance contract with Automotive Management Services was still in effect in 2015. A continuing obstacle was that the contractor services only those vehicles with a registered Vehicle Identification Number. The ALP SOAG said it was working to register new vehicles, as well as any older vehicles that had been missed, and

---

[76] MOI official, statements at a conference attended by the authors, February 13, 2013.

[77] NSOCC-A, 2013i.

[78] MOI official, statements at a conference attended by the authors, February 13, 2013.

[79] Regional Command East officials, interviews with the authors, February 13, 2013; and NTM-A officials, interviews with the authors, February 17, and 24, 2013.

[80] Regional Command East official, interview with the authors, February 13, 2013.

[81] Regional Command East official, interview with the authors, February 13, 2013.

[82] MOI official, interview with the authors, February 21, 2013.

did not expect any issues after that effort was complete. Furthermore, one SOAG adviser stated that he was unaware of any complaints about vehicle maintenance from ALP headquarters.[83] However, another adviser noted that ALP units in the field were very reluctant to bring their vehicles to regional maintenance centers, as each movement represented a security risk and loss of manpower for the unit.[84]

The maintenance of weapons is also problematic for MOI. Many ALP units are operating with VX-58 and AMD-65 rifles, both considered of lower quality than the Russian AK-47s used by other Afghan police units. These weapons were acquired without sustainment packages for the units receiving them. As a result, one Coalition official estimated in February 2015 that 33 percent of the current ALP weapons were serviceable.[85] Three options to address this issue were being considered. First, MOI or the Ministry of Defense could transfer spare parts in their inventory to keep these weapons serviceable. While seemingly the simplest option, the bureaucratic hurdles were formidable, and it was unclear whether sufficient repair parts could be obtained through this method. Second, the Afghans could seek another donor, such as China, for weapons that come with a sustainment package. Finally, the United States could purchase weapons for ALP and then service them through U.S. sources.[86] However, one interviewee felt that the higher maintenance and proficiency standards needed to keep a U.S. weapon working made this option impractical for ALP.[87]

Checkpoint Facilities

As mostly part-time home-based forces, ALP guardians have no use for large government facilities; the one exception is the national headquarters building in Kabul. The only other facilities that ALP personnel operate are the defensive checkpoints surrounding their villages.

---

[83] NSOCC-A official, interview with the authors, February 25, 2015.

[84] NSOCC-A official, interview with the authors, February 25, 2015.

[85] CSTC-A official, interview with the authors, February 25, 2015.

[86] NSOCC-A official, interview with the authors, February 25, 2015.

[87] NSOCC-A official, interview with the authors, February 26, 2015.

According to one Coalition contractor we spoke to in 2013, the checkpoints manned by ALP and other elements of ANP were in generally poor condition, which had an effect on these forces' morale and rate of attrition.[88] Other Coalition officials found existing ALP checkpoints to be Spartan yet "good enough," meaning they were designed to provide only minimal shelter and protection.[89] The more significant problem was that the rapid expansion of the ALP program had made it more difficult to provide a sufficient number of checkpoints in newly validated districts.[90] To assist in the supply of checkpoint materials and their establishment, the Coalition had begun providing "checkpoint kits." These kits included sandbags, protective barriers, rails, and other items. As of February 2015, 80 checkpoint kits had been distributed to seven RLCs—enough for roughly 240 checkpoints. Reportedly, this allocation was the result of a deliberate process of determining checkpoint requirements in each ALP district; thus, not every RLC has received the same number of kits.[91]

Medical Support

By Western standards, medical care in Afghanistan is not very good, and it is particularly rudimentary in the rural parts of the country where most ALP guardians reside.[92] Until mid-2013, ALP units could rely on their SOF team mentors and their aerial medical evacuation capability when their personnel were seriously injured and required transportation to either a Coalition facility or a local hospital.[93] But that capability was withdrawn when Coalition forces were reduced and ALP districts transitioned to Afghan control. From that point on, ALP has had to rely primarily on AUP or ANA for medical assistance,

---

[88] NTM-A contractor, interview with the authors, February 20, 2013.

[89] NTM-A officials, interviews with the authors, February 21 and 22, 2013.

[90] NTM-A official, interview with the authors, February 21, 2013.

[91] NSOCC-A official, interview with the authors, February 25, 2015.

[92] NTM-A official, interview with the authors, February 17, 2013.

[93] SOF team member, interview with the authors, February 2013; and NSOCC-A official, interview with the authors, February 15, 2013.

which includes a casualty evacuation capability but not medical evacuation.[94] When interviewed in 2013, one Coalition expert stated that while ALP medical support was improving,[95] "the best hope is that ANP medical capacity won't deteriorate once the Coalition leaves."[96] The results as of winter 2015 were not promising. ALP did acquire a medical funding line in its budget, but ALP headquarters could not determine what could be purchased with these funds. Unable to resolve the issue, the money had to be returned to MOI at the expiration of the budget cycle.[97] Nonetheless, ALP could improve its own medical capabilities. For example, MOI in 2013 proposed an allocation of one medic, trained by the surgeon general's office, to every ten ALP guardians.[98] Although this may have been wishful thinking, DynCorps was given a contract for medical training to ALP, which could supplement the basic first aid training offered by the Red Cross.[99] Unfortunately, the study team was unable to obtain detailed information on the medical support being provided to ALP in 2015.

Communications

ALP personnel in the field manage with very high-frequency radios and cell phones to communicate with one another and the district headquarters, but communications between ALP headquarters and the district headquarters have been problematic. According to one SOF mentoring team in 2013, its ALP personnel rely mostly on cell phones for communication support, including calling for Quick Reaction Force assistance. While this seemed to work well enough for those ALP personnel, they complained about not being compensated for the

---

[94] SOF team member, interview with the authors, February 17, 2013.

[95] NSOCC-A official, interview with the authors, February 16, 2013.

[96] NTM-A official, interview with the authors, February 17, 2013.

[97] NSOCC-A official, interview with the authors, February 25, 2015.

[98] MOI officials, interviews with the authors, February 19, 2013; and NTM-A officials, interviews with the authors, February 17 and 21, 2013.

[99] SOF team member, interview with the authors, February 24, 2013; and NTM-A official, interview with the authors, February 17, 2013.

minutes they used on their personal cell phones.[100] Communications from the national to district levels are in part hampered by infrastructure limitations. MOI support for communication systems and information technology stops at the provincial headquarters,[101] and one regional Afghan military official interviewed in 2013 stated that even at that level, systems were sometimes inadequate.[102] Communication difficulties at the district level are compounded by bureaucratic politics. According to one ALP headquarters official in 2013, ALP commanders were not permitted to use the communication equipment that belonged to DCOPs.[103]

Conclusion

MOI has made significant advances in improving its logistics practices and results, yet much more needs to be done if it is to be able to effectively manage its logistics system on its own. The Coalition's 2012 initiative to eliminate bottlenecks in the process of providing ALP with initial equipment was mostly successful, and its chances for continued success under full Afghan control are relatively good, particularly if RTCs can continue to be used as the main venue for supplying new guardians with their initial kits. However, the resupply situation is more problematic, largely because of provincial and district reluctance to provide ALP with substantial quantities of supplies. Other ALP-related logistics problems include the loss of Coalition air transport, limited provincial and district storage facilities, inadequate vehicle maintenance education and support, limited medical capabilities and support, and unreliable communications between ALP headquarters and the district headquarters.

---

[100] SOF team member, interview with the authors, February 2013.

[101] NTM-A official, interview with the authors, February 17, 2013.

[102] Regional Command East official, interview with the authors, February 13, 2013.

[103] MOI official, interview with the authors, February 19, 2013.

CHAPTER FOUR

Personnel Management

Afghans have begun to acquire and demonstrate many of the capabilities necessary to successfully manage ALP personnel. For example, Afghan elders, government officials, and contractors in early 2015 were handling all ALP recruiting, vetting, and in-processing tasks with no assistance from the Coalition. Furthermore, MOI was, for the most part, paying ALP personnel in a timely manner—although many Afghans groused that the amount of ALP pay and benefits was insufficient. Despite these mostly hopeful signs, it is not yet clear whether the Afghans can maintain the current level of personnel management capability once Coalition mentors and enablers are removed from the equation. One indicator of managerial weakness is MOI's inability to authorize and fill many of the positions needed to support ALP guardians and other members of ANP.

MOI Administrative Capacity

ALP headquarters has a limited capacity to oversee the approximately 29,000 guardians who operate at the village level throughout much of Afghanistan. In its tashkil authorizations, it has no personnel who work directly for the headquarters at provincial and district levels. Instead, MOI officials reporting primarily to local governors and chiefs of police administer critical support functions for ALP, as they do for other elements of ANP.[1] In 2013, one NSOCC-A adviser lamented

[1] SOF team member, interview with the authors, February 2013; and MOI official, interview with the authors, February 19, 2013.

the lack of an ALP representative to relay problems up the chain of command.[2] Two years later, another adviser indicated that there was a reporting system, albeit slower than he would prefer, to bring attention to ALP issues in the districts.[3] Another issue raised in 2013 was that MOI had not put sufficient effort into creating new management positions, or even filling authorized managerial positions, to cope with the support requirements of a still-growing ALP force.[4] As of February 2015, 85 percent of AUP officer positions authorized by the tashkil for ALP district support nationwide were filled (261 of 307), while only 70 percent of noncommissioned officer positions were filled (198 of 283).[5] This is a significant improvement over 2013, when only 50 percent of ALP support positions were filled.[6]

Unable to retain trained personnel, manpower at the national warehouses was decreasing in 2013, yet requests for equipment were on the rise, leading to delays in the supply process.[7] Manpower was reportedly tight at RLCs as well.[8] Furthermore, although the ALP headquarters in 2013 hoped to place representatives in the RLCs to facilitate the push of ALP equipment to lower levels, this desire had not been realized.[9] Arguably, the need for such ALP liaison billets was even greater in provincial police headquarters, where much of the hoarding and diversion of ALP supplies was taking place. Senior ALP officials and their Coalition advisers once favored this addition to the provincial tashkil, although the efficacy of this organizational reform is unclear, given that the AUP officer in question would be under the PCOP's

---

[2] NSOCC-A official, interview with the authors, February 13, 2013.

[3] NSOCC-A official, interview with the authors, February 25, 2015.

[4] NSOCC-A official, interview with the authors, February 17, 2013.

[5] NSOCC-A, "ALP Strength Report," Kabul, Afghanistan, February 21, 2015a.

[6] NTM-A official, interview with the authors, February 22, 2013.

[7] MOI officials, interviews with the authors, February 14 and 25, 2013.

[8] MOI official, interview with the authors, February 14, 2013.

[9] NSOCC-A officials, interviews with the authors, February 17 and 25, 2013; and NTM-A officials, interviews with the authors, February 17 and 25, 2013.

command.[10] In 2013, districts had a tashkil that included one AUP officer and two noncommissioned officers devoted to ALP management, communications, and logistics.[11] Despite their critical importance to MOI's ability to support ALP, however, there was still a shortage of logistics representatives in ALP districts.[12]

Coalition officials in 2013 and 2015 raised questions about MOI's capability to support ALP regardless of the level of tashkil authorizations. National MOI leaders, they felt, were ignorant of, and detached from, the particular needs and problems of ALP.[13] According to one 2015 Coalition adviser, the prevailing MOI attitude was that "the Americans created ALP, so the Americans can support them."[14] Also, communication and accountability mechanisms connecting the national headquarters to ALP in the field continue to evolve. The ALP headquarters personnel manager has lobbied for better systems and greater accountability, particularly in the areas of recruitment and martyrdom/disability payments.[15] Additionally, in 2015, the finance and contracts team began "spot checking" ALP provincial reporting.[16] ALP was initially left out of a larger effort to establish automated administrative systems within MOI, such as electronic fund transfer for distributing police salaries, because of its lack of access to Law and Order Trust Fund Afghanistan, which funds most of ANP.[17] While

---

[10] Regional Command East officials, interviews with the authors, February 13, 2013; and NSOCC-A officials, interviews with the authors, February 13 and 17, 2013.

[11] NTM-A official, interview with the authors, February 22, 2013.

[12] NSOCC-A officials, interviews with the authors, February 14 and 17, 2013; and NTM-A officials, interviews with the authors, February 17, 2013.

[13] NTM-A official, interview with the authors, February 22, 2013.

[14] NSOCC-A official, interview with the authors, February 25, 2015.

[15] NSOCC-A official, interview with the authors, February 25, 2015.

[16] As of February 2015, the ALP headquarters had requested officials in two provinces to send time and attendance reports in order to search for personnel irregularities. While this proactivity was encouraging, Coalition advisers reported that their financial officers can handle only two provinces at a time, making a countrywide audit a time-consuming process. (NSOCC-A official, interview with the authors, February 25, 2015)

[17] NSOCC-A official, interview with the authors, February 25, 2015.

some progress has been made on establishing computerized reporting for ALP, most reporting is still paper-based. As discussed earlier, some advisers indicated that the existing system was slow but effective. But the disparity in managerial mechanisms between ALP and other MOI institutions complicates the plan to integrate ALP with the rest of the police force. In addition to the systems themselves, Coalition efforts to create a cadre of computer-literate administrators who could digitally track ALP personnel and logistics actions seem to have stalled.[18] One Coalition adviser to MOI interviewed in 2015 said that it would take years for the ministry to develop the administrative capacity necessary to oversee ALP.[19]

ALP Recruiting, Vetting, and In-Processing

Most MOI and Coalition officials we interviewed in 2013 expressed satisfaction with ALP recruiting, vetting, and in-processing procedures. One senior MOI officer said that these procedures surpassed those used by other elements of ANP and ANA.[20] While some also were confident that the Afghans could perform the "due diligence necessary for recruiting and vetting of ALP" without Coalition oversight, others felt that MOI would be unable to maintain tribal balances within ALP units and keep at bay criminal, insurgent, or powerbroker influences in the long term.[21] See Table 4.1 for an outline of the ALP tribal/ethnic breakdown compared with the breakdown in Afghanistan as a whole.

Recruiting

The SOF-inspired practices for selecting and recruiting ALP guardians are enshrined in MOI doctrine. According to the 2013 ALP establishment procedures, ALP personnel must come from and be endorsed by

[18] NSOCC-A official, interview with the authors, February 25, 2013.

[19] CSTC-A official, interview with the authors, February 25, 2015.

[20] MOI official, interview with the authors, February 14, 2013.

[21] MOI officials, interviews with the authors, February 14 and 25, 2013.

Table 4.1
Tribal/Ethnic Breakdown Compared to Afghanistan as a Whole

Tribe/Ethnicity	ALP Areas (%)	Afghanistan (%)
Pashtun	73.0	42.0
Tajik	11.0	31.0
Uzbek	9.0	9.0
Hazara	3.0	10.0
Aimak	1.0	2.0
Turkmen	1.0	2.0
Baluch	0.2	1.0
Others (e.g., Arab, Nuristani)	1.0	3.0

SOURCE: The numbers for ALP areas are based on surveying in 80 districts near known ALP locations in March 2013. The data for Afghanistan are estimates based on Asia Foundation, *A Survey of the Afghan People*, Washington, D.C., November 17, 2015. No census has been conducted in Afghanistan since the 1970s.

the communities they serve. Doctrinally, in order to be considered for the program, prospective ALP recruits must meet the following criteria:

- Be patriotic, disciplined, and faithful to GIRoA development and prosperity.
- Receive an original, valid national identification document (Tazkera).
- Meet physical and mental health criteria and not be addicted to any drugs or alcohol.
- Be willing to serve for a three-year period; contracts can be extended based on the need and approval of the ALP Directorate.
- Be at least 18 and not more than 45 years of age (however, where there are not enough qualified recruits, an age waiver up to 50 years may be allowed by MOI).
- Sign or thumb a written copy of a service commitment document about their assignments.
- Submit to background checks of their living place; recruits should be verified and endorsed by local intelligence organizations, such as anticrime and antiterrorism organizations.

- Preferably be literate; educated volunteers will be given priority to join ALP.
- Have a good reputation in the society, have no criminal record, and not be convicted of any crime of human rights.
- Fill out and provide medical forms to verify health status.
- Obey the prevailing laws of the country.
- Protect public property and safeguard their weapons and equipment.
- Carry their MOI-issued ALP identification card.[22]

The fulfillment of these requirements is documented in an application packet that approaches 18 pages when filled out. This application includes character references and is checked by local and national police agencies before a candidate may proceed to in-processing.[23]

According to the ALP SOAG, the practices of village elders nominating ALP guardians and police review of their applications continued in 2015.[24] However, nongovernmental organizations active in Afghanistan have complained for some time that ALP recruitment standards have not been met in some parts of the country. Back in 2011, for example, Human Rights Watch indicated that a local strongman in the northern province of Baghlan had recruited into ALP former fighters with the Islamist Hezb-i-Islam, and these fighters had become implicated in various illegal activities.[25] Four years later, in the wake of the Taliban's temporary takeover of the city of Kunduz, the Afghan Analysts Network quoted a provincial police commander in Baghlan as saying that more than 700 young men had registered for service in local militias who did not appear on MOI's tashkil.[26]

---

[22] NSOCC-A, *Afghan Local Police In-Processing Enrollment Handbook*, Kabul, Afghanistan, May 2013h.

[23] NSOCC-A official, interview with the authors, February 25, 2015.

[24] NSOCC-A official, interview with the authors, February 25, 2015.

[25] Human Rights Watch, "Afghanistan: Rein in Abusive Militias and Afghan Local Police," Kabul, Afghanistan, September 12, 2011.

[26] Gran Hewad, "The 2015 Insurgency in the North (4): Surrounding the Cities in Baghlan," Afghanistan Analysts Network, October 21, 2015.

Vetting and In-Processing

In 2013, the processing of ALP guardians was carried out at the district level and at RTCs by specially tailored in-processing teams composed of Afghan contractors and NSOCC-A advisers.[27] While Afghans largely conducted most of the in-processing tasks, they relied on Coalition forces for certain requirements, including transportation and security.[28] Also, because the Coalition biometric database is far more extensive than that of MOI, Afghan contractors often ran potential ALP candidates through the Coalition system, as well as through GIRoA's.[29] But this dependence came to an end in June 2013, when NSOCC-A personnel discontinued their direct role in processing ALP candidates, although they continued to offer guidance to MOI as necessary.[30] To facilitate this transition, NSOCC-A published an *Afghan Local Police In-Processing Enrollment Handbook*.[31] Coalition advisers in 2015 observed that MOI now has processes in place for accessing ALP guardians that are in line with this guidance.[32]

The guide, released in May 2013, notes that MOI has the ability to in-process ALP at RTCs, provincial training or recruiting centers, and certain recruit collection points. In-processing occurs in two stages:

> First, the recruiter completes the ALP contract and identification card application with each selected candidate. The recruiter ensures all required signatures (from the local Shura leader/ District Chief of Police) and endorsements are completed for this paperwork. . . . For ALP, the recruiter also functions as the personnel officer at the district level and is responsible for keeping personnel records for each ALP member.[33]

---

[27] NSOCC-A official, interview with the authors, February 14, 2013.

[28] NSOCC-A official, interview with the authors, February 22, 2013.

[29] SOF team member, interview with the authors, February 24, 2013.

[30] NSOCC-A, 2013e.

[31] NSOCC-A, 2013h; and NSOCC-A, "ALP Weekly Update," Kabul, Afghanistan, May 20, 2013j.

[32] NSOCC-A official, interview with the authors, February 25, 2015.

[33] NSOCC-A, 2013h.

Provincial recruitment officers, along with DCOPs, are responsible for ensuring that the ALP contract, M-41 pay document, and background check documents are completed before recruits are taken to recruit collection points. The PCOP or DCOP provides transportation to a training location if necessary.[34] In order to facilitate in-processing at more-remote locations, the ANP General Recruiting Command supports and provides select provincial recruiting officers with biometric kits and urine-analysis kits.[35]

Stage two involves enrollment into the MOI system. GIRoA biometrics collection teams and identification card teams in-process recruits on-site. These teams collect biometrics on each candidate in order to build the Afghan Automated Biometric Identification System. The identification card applications are then collected by the identification card team and returned to Kabul, where they are submitted to the ANP General Recruiting Command for approval. Once signed, the MOI identification card office manufactures the cards. At the same time, ALP members are enrolled in the MOI personnel and finance systems.[36]

The effectiveness of these vetting and in-processing procedures was called into question in 2012 by a rash of so-called green-on-blue attacks by Afghan security forces, including ALP, on Coalition personnel. As a result, NSOCC-A and MOI initiated a plan to re-vet every guardian in the ALP program.[37] By the end of January 2013, 100 percent of ALP personnel had been re-vetted, and NSOCC-A had committed itself to helping MOI reinforce its recruiting and vetting procedures.[38] As of February 2015, Coalition advisers reported that they were satisfied with what they knew of MOI's implementation of these procedures.[39]

---

[34] NSOCC-A official, interview with the authors, February 25, 2015.

[35] NSOCC-A, 2013h.

[36] NSOCC-A, 2013h.

[37] New America Foundation, "Attacks on U.S. and NATO Soldiers by Afghan Security Forces," web page, October 18, 2012.

[38] NSOCC-A, "ALP Weekly Update," Kabul, Afghanistan, March 1, 2013c.

[39] NSOCC-A official, interview with the authors, February 25, 2015.

Recontracting

Once ALP contracts expire, guardians and ALP members have the option of reenlisting. Provincial recruitment officers are responsible for recontracting, with assistance from the DCOP or PCOP. Reenlisting tasks are similar to the original in-processing procedure, including the requirement to be re-vetted by the local shura. During recontracting, however, guardians are not required to biometrically enroll again and are required to renew their identification card application only if their card was never issued during their first in-processing.[40]

Personnel Results

According to the study team's analysis of ALP data from January 2012 to January 2013, districts that had recently transitioned to Afghan control for at least three months while still being in SOF tactical overwatch were maintaining their authorized levels of personnel. Although individual district results varied, the median aggregate monthly personnel level for these districts was nearly 100 percent. In contrast, the level was less than 30 percent in the non–tactical overwatch districts that we examined, where the SOF push to bring ALP up to standard was not yet under way.[41]

Despite suffering heavy casualties, ALP personnel levels appeared to be remarkably high in 2015. Some researchers have attributed this to rising unemployment in the country, as well as greater insecurity in rural areas, which has induced anti-Taliban villagers to arm themselves against the insurgency.[42] NSOCC-A estimated ALP retention at 93 percent and monthly attrition at 1–2 percent in the spring of 2015.[43]

---

[40] NSOCC-A, 2013h.

[41] NSOCC-A, ALP data provided to RAND, January 2012–January 2013.

[42] International Crisis Group, 2015.

[43] Special Inspector General for Afghanistan Reconstruction, *Quarterly Report to the United States Congress*, Arlington, Va., April 30, 2015a, p. 95.

ALP Pay and Benefits

The effectiveness and integrity of the ALP pay system varies greatly across Afghanistan. As of February 2015, ALP guardians were supposed to be paid in accordance with MOI's ALP pay process shown in Figure 4.1. Pay and benefits have been a perennial issue across the force, however. Often, the source of the problem has been either a budget not being passed in a timely manner or poor administrative practices.[44] Unable to influence the budget timeline, ALP headquarters and its Coalition advisers have focused on developing processes to make the ALP pay system more efficient, timely, and fair. For example, MOI has made progress in converting all ALP guardians over to electronic fund transfer payments, with 42 percent on such payments as of May 2015.[45] It is hoped that this method of payment will reduce opportunities for corruption and delay that are inherent in the still-dominant "trusted agent" delivery mechanism.

That said, MOI officials in 2013 complained that ALP guardians' pay and benefits were insufficient for the vital and dangerous work they were performing.[46] ALP personnel receive a base pay according to their rank, along with a monthly food stipend. In 2015, this worked out to a monthly net salary of 6,000 Afghanis (approximately $93) for the average ALP guardian, 7,250 Afghanis (approximately $113) for the ALP group leader, and 8,250 Afghanis (approximately $128) for the ALP team leader.[47]

Although the ALP base salary is less than that of counterparts in the rest of ANP, this disparity was reduced in 2013, when the ALP food stipend increased to 3,750 from 3,250 Afghanis, bringing the overall pay package of the average ALP guardian in line with that of

---

[44] NSOCC-A official, interview with the authors, February 25, 2015.

[45] NSOCC-A, "ALP Introductory Brief," Kabul, Afghanistan, February 25, 2015b.

[46] MOI officials, interviews with the authors, February 18 and 19, 2013.

[47] These salaries are somewhat higher than those received by many lower-skilled Afghans, such as farmers and fisherman, although considerably less than the median salary in most other professions, including construction.

Figure 4.1
Current Afghan Local Police Eligibility and Pay Process

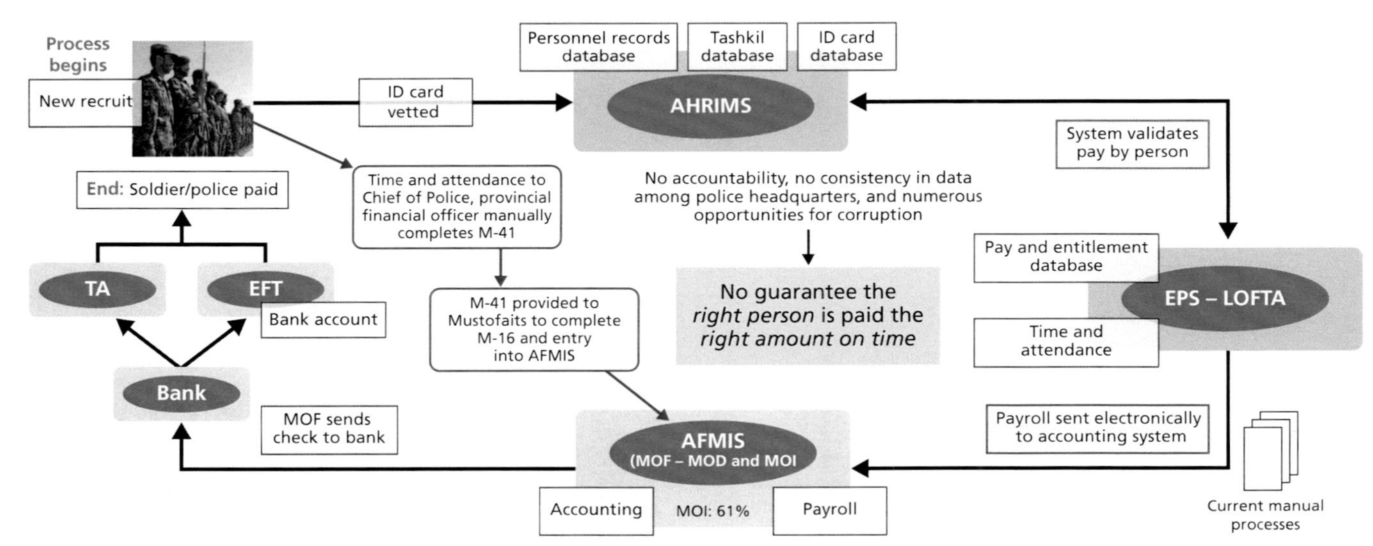

SOURCE: International Security Assistance Force, "Personnel and Payroll System Relationships," presentation, Kabul, Afghanistan, December 30, 2014. Photo: U.S. Department of Defense.

NOTE: The red boxes mean that those databases are not being tapped in the current ALP eligibility and pay process. AFMIS = Afghanistan Financial Management Information System; AHRIMS = Afghan Human Resource Information Management System; EFT = electronic fund transfer; EPS = electronic payroll system; LOTFA = Law and Order Trust Fund for Afghanistan; MOD = Ministry of Defense; MOF = Ministry of Finance; TA = trusted agent.

RAND *RR1399-4.1*

a junior AUP officer, who is not accorded a food stipend.[48] However, AUP officers have more opportunities for promotion and attendant salary increases throughout their career, whereas the position of ALP guardian is mostly dead-end (although ALP veterans are given preferences for hiring within the regular police force and the army). Also, in contrast to ANA and other ANP elements, ALP personnel do not receive hazard pay, despite the fact that they come into contact with insurgents more frequently than do ANA and ANP personnel.[49]

Payment Processes

Although the prevalence of the payment method depends on the region, the majority of ALP guardians receive their salary and food stipend from so-called trusted agents, either monthly or bimonthly. Under this system, MOI provides the payroll in cash to an individual, who then travels to ALP districts and pays individual ALP guardians. This system has the benefit of being able to reach ALP personnel operating in remote areas with few local banks. Advisers identified two disadvantages to this approach. The first is its susceptibility to corruption (although some argued that the trusted agent system provides accountability by requiring ALP guardians to appear in person with their guns in order to receive their pay).[50] The second drawback is that many of the ALP guardians are illiterate. They must trust someone to explain their contract, tell them how much they are owed, and provide their pay; this agent is designated and "trusted" by MOI, not necessarily known to the local ALP personnel. Hence, there are many opportunities for the trusted agents to take advantage of the process.[51]

As noted previously, 42 percent of ALP guardians receive their salary and food stipend through an electronic payroll system. Used extensively throughout MOI and the Ministry of Defense, this system

---

[48] NTM-A official, interview with the authors, February 22, 2013.

[49] NTMA-A official, interview with the authors, February 20, 2013.

[50] NSOCC-A official, interview with the authors, February 14, 2013.

[51] NSOCC-A official, interview with the authors, February 25, 2015.

transfers ALP pay to individual bank accounts.[52] This method has the advantage of improving ALP salary accountability but is unsuitable for many Afghans who do not have bank accounts or regular access to banks.[53] MOI has issued a cipher that mandates 100-percent enrollment in electronic fund transfers, but Coalition advisers we interviewed in 2013 believe that this is not practical—especially for guardians in remote areas without consistent access to electronic funds. Instead, advisers anticipated that approximately 70 percent of the force could be enrolled in electronic fund transfers.[54]

Despite the necessity of continuing to use trusted agents in some cases, Coalition advisers and their MOI counterparts have continued to develop better systems to account for and pay ALP guardians. Figure 4.2 shows a proposed system that Coalition advisers hope will be implemented in 2016.[55] In addition to standardizing ALP pay and reducing corruption, this system will also provide a tie-in to the Afghan Human Resources Information Management System. Advisers hope that the verification of a guardian's status in the system will reduce instances of fraudulent submissions and better account for casualties, desertions, and those who leave ALP.

One glitch in the overall ALP payment process is that MOI policy restricts ALP recruits from being paid until they have fully completed in-processing and training, leaving new guardians without their salaries for at least three months. Moreover, many ALP guardians experience a pay delay of several weeks to several months following the end of their training period. In 2013, this led the Coalition to step in to fill the pay gap with money from the Afghanistan Security Forces Fund until MOI payments arrived.[56] However, this was just a temporary measure; as of July 2013, Coalition forces were no longer authorized to pay ALP salaries using this funding, putting pressure on district

---

[52] NSOCC-A official, interview with the authors, February 16, 2013.

[53] SOF team member, interview with the authors, February 24, 2013.

[54] NSOCC-A official, interview with the authors, February 25, 2015.

[55] NSOCC-A official, interview with the authors, February 25, 2015.

[56] NSOCC-A, 2013e.

Figure 4.2
Proposed Afghan Local Police Eligibility and Pay Process

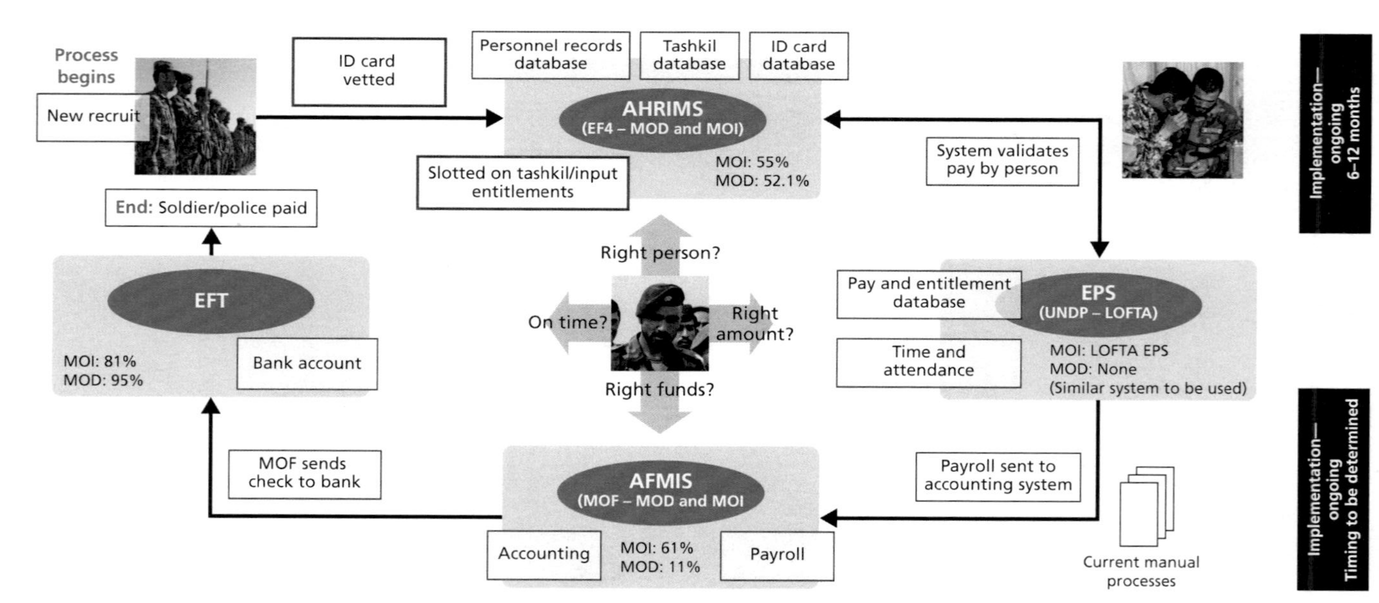

SOURCE: International Security Assistance Force, 2014. Photos: U.S. Department of Defense; U.S. Marines; SGT Mick Davis.
NOTE: AFMIS = Afghanistan Financial Management Information System; AHRIMS = Afghan Human Resource Information Management System; EFT = electronic fund transfer; EPS = electronic payroll system; LOTFA = Law and Order Trust Fund for Afghanistan; MOD = Ministry of Defense; MOF = Ministry of Finance; UNDP = United Nations Development Program.
RAND *RR1399-4.2*

and provincial authorities to get ALP candidates on the MOI payroll as soon as possible.[57] The expectation was that in-processing ALP personnel at RTCs would ease the entry of new guardians into the MOI payment system.[58] But the initial payment problem raised a larger issue about whether local officials would continue to fill out the paperwork necessary to pay their ALP forces without assistance and encouragement from Coalition mentors.[59]

Early in the transition period, the pay situation seemed to be improving. For example, the distribution of paychecks to "ghost" ALP—former guardians who have been fired or left the force—appeared to be less frequent than it had been. One former SOF team leader we interviewed noted that 30 to 40 such former ALP personnel were getting paid when he first arrived in his district, and at the time of our interview, it was five to ten.[60] This was a consequence, he believed, of the provincial headquarters' willingness to purge its pay logs of fraudulent entries.[61]

Subsequently, however, there were signs that the pay system was not functioning properly for at least a portion of the ALP force. According to a 2015 report by the Special Inspector General for Afghanistan Reconstruction, Coalition officials said they received "frequent reports of ALP personnel not receiving their full pay." In addition, the audit of payroll records from at least five different provinces over at least five months within the previous year turned up "several irregularities, primarily with the data collected and the forms used to facilitate the salary disbursement process."[62]

---

[57] NSOCC-A, "ALP Weekly Update," Kabul, Afghanistan, April 15, 2013f.

[58] NSOCC-A official, interview with the author, February 14, 2013.

[59] NSOCC-A officials, interviews with the authors, February 14 and 17, 2013.

[60] SOF team member, interview with the authors, February 2013.

[61] SOF team member, interview with the authors, February 2013.

[62] Special Inspector General for Afghanistan Reconstruction, 2015.

Pay Sufficiency

Despite the increase in ALP's food stipend, Coalition and Afghan officials in 2013 expressed different views regarding the sufficiency of ALP pay. To one senior MOI leader, the relatively small salary for ALP personnel was a "big problem."[63] This claim that ALP forces deserved an unspecified boost in salary was based, in part, on the role the guardians played as first-line defenders against insurgents throughout the country, as well as on the disproportionate sacrifices they were making in terms of casualties. The interviewee also argued that pay and benefit inequality was the source of friction between ALP and the rest of ANP.[64] Some concluded from this that ALP should receive the same pay, benefits, and training as AUP.[65] Others thought that they should be paid the same as AUP when they performed similar duties as AUP did, such as manning checkpoints at a substantial distance from their villages; however, they should not necessarily receive additional compensation when they remained in their villages and pursued their normal activities.[66]

One Coalition adviser we spoke to in 2013 agreed that ALP salaries should be augmented to some extent, although he wondered where the additional funds would come from. But other members of the Coalition had less sympathy with the notion that ALP forces deserved a significant pay increase, noting that ALP guardians had one important benefit that ANP and AUP did not: They could remain in their home villages and did not have to serve in a faraway region separated from their families. Furthermore, ALP guardians could apply to join the army or other elements of the police force if they so chose. Finally, they found it unrealistic to consider raising ALP salaries to the level of AUP and ANA, when it seems likely that GIRoA will not have the resources to pay for either component of the Afghan National Security

---

[63] MOI official, interview with the authors, February 21, 2013.

[64] MOI officials, interviews with authors, February 18 and 19, 2013.

[65] MOI official, interview with the authors, February 19, 2013.

[66] MOI official, interview with the authors, February 21, 2013.

Forces without substantially reducing their end strengths.[67] Furthermore, as one Coalition adviser pointed out in 2015, ALP was then manned to more than 95 percent of its tashkil appropriation, and there did not appear to be a problem getting and keeping guardians at the existing pay rate.[68]

Martyrdom Benefits

Martyrdom benefits are particularly important for a force whose members are often from poorer elements of society and whose casualty rates are higher than their counterparts in other services. The Afghan government has the complete responsibility for martyrdom payments, thus leaving little role for Coalition advisers—although the topic is frequently discussed.[69] At ALP headquarters, a personnel official in 2015 indicated that martyrdom and disability were second on his list of priority issues after recruitment.[70] Despite the Afghan government's provision of martyrdom payments to family members of deceased in all the security forces, there is one difference in the process for relatives of ALP guardians. ALP family members must come to Kabul to complete the required paperwork. This is problematic for those who need to travel long distances to an unfamiliar city and navigate the MOI bureaucracy, especially when the cost of the trip must be borne by the family without knowing if its request will be successful.[71] Therefore, it is essential that MOI and ALP headquarters develop a fairer and less onerous method for applying for and receiving the martyrdom benefits of fallen members of the ALP force.

---

[67] NSOCC-A official, interview with the authors, February 22, 2013; and NTM-A official, interview with the authors, February 22, 2013.

[68] NSOCC-A official, interview with the authors, February 25, 2015.

[69] NSOCC-A official, interview with the authors, February 25, 2015.

[70] NSOCC-A official, interview with the authors, February 25, 2015.

[71] NSOCC-A official, interview with the authors, February 25, 2015.

Conclusion

MOI is beginning to develop effective and sustainable personnel processes for the administration of ALP. Although there have been recent signs of deterioration in the distribution system, most ALP guardians were getting paid in a timely manner in 2013, and the spread of electronic payment methods to ALP has the potential to reduce corruption and delay in the pay system. On the downside, MOI has a limited tashkil authorization to support ALP and other elements of ANP, and some doubt MOI's ability to maintain tribal balances within ALP units or prevent their takeover by other local men without the consent of village leaders. Furthermore, it is unclear whether MOI will have the ability to independently vet ALP guardians and conduct the full range of in-processing tasks in the future. For their part, MOI officials have complained that ALP guardian pay is insufficient for the important work they do and a source of tension between ALP and other elements of ANP.

CHAPTER FIVE

Training

The state of ALP training is good compared with the situation in the rest of the Afghan police force. In early 2015, approximately 86 percent of the ALP force had attended a formal training course, with 14 percent remaining temporarily untrained, mostly because security threats prevent them from leaving their villages.[1] Furthermore, the regional training concept introduced by the Coalition in late 2012 appears to be an essential component of the long-term training solution for ALP following the departure of the last remaining village-based Coalition SOF trainers and mentors in 2014. The RTCs selected for ALP training, many of which are colocated with RLCs, provide a one-stop opportunity for ALP in-processing, equipping, and training. Although not without start-up problems, the RTCs and their AUP trainers have demonstrated their ability to meet the demand for trained ALP recruits during a period of significant attrition for the force.[2]

Training Location

From the ALP program's inception until late 2012, training revolved around embedded Coalition SOF teams providing instruction to ALP recruits at the village level, often in cooperation with AUP units under the leadership of DCOPs. On-site training was in line with program

[1] NSOCC-A official, interview with the authors, February 25, 2015.

[2] SOF team members, interviews with the authors, February 2013 and February 17, 2013.

doctrine, which emphasized local buy-in and ownership. SOF teams could recruit and vet potential guardians, in-process them, and begin training them in their communities in relatively short order. Meanwhile, guardians could remain in their homes while providing protection to fellow villagers and situational awareness to their SOF mentors.

In 2013, NSOCC-A encouraged MOI to open its provincial- and regional-level training facilities to ALP personnel. In the words of an NSOCC-A official publication,

> a new model will be necessary for training and sustaining the ALP. . . . With the 2014 transition to [Afghan National Security Forces] lead and as GIRoA security capacity builds, GIRoA will increasingly have to be able to generate new ALP formations and regenerate new guardians with minimal NSOCC-A support.[3]

And constructing that independent capability within MOI required the bulk of ALP training to shift to centralized locations. However, Coalition and MOI leaders recognized that local training efforts also needed support. Consequently, the 2013 ALP establishment procedures called on ALP program managers to cooperate with MOI's General Training and Education Command "in order to provide rotational training for ALP personnel close to their duty stations through mobile training teams."[4]

The consensus among Coalition advisers in 2013 was that a hybrid training system, with local and regional aspects, was the only realistic option.[5] The location of training—whether at the regional, provincial, district, or village level—should depend on local conditions, they believed. Because the ease or difficulty of transporting equipment and personnel from ALP sites to potential training locations varies considerably, the selection of actual training locations also should vary.[6]

---

[3] NSOCC-A, "ALP Weekly Update," Kabul, Afghanistan, January 21, 2013a.

[4] Islamic Republic of Afghanistan, 2013a.

[5] NSOCC-A officials, interviews with the authors, February 22 and 25, 2013; and NTM-A officials, interviews with the authors, February 22 and 25, 2013.

[6] NSOCC-A official, interview with the authors, February 14, 2013.

Regional Training Centers

As of August 2014, training of ALP guardians was being conducted at 11 RTCs across Afghanistan, with transport of students to the RTCs being handled by their PCOP or DCOP.[7] The training conducted at the RTCs is entirely Afghan-led; there is no longer any Coalition involvement in ALP training.[8] On occasion, when ALP guardians are unable to travel to an RTC, their training has been conducted by ANA Special Forces teams. While this is not a common occurrence, it is an option for MOI when travel is impractical.[9]

Strategy

According to NSOCC-A, the regional and provincial training system is an "Afghan sustainable framework for the implementation of Afghan-led formal ALP training at designated [sites]."[10] As indicated above, the goals of this centralized system when it was developed were to establish an enduring, Afghan-sustainable training capability and to set the conditions for SOF transition to operational-level advising by the end of 2013.[11] Most of the RTCs designated for ALP are preexisting ANP training facilities, complete with all the facilities necessary to provide basic training to ALP recruits.[12] The DCOP for each ALP formation coordinates the movement of his ALP candidates to the RTC, where AUP trainers provide standardized, hands-on instruction based on the existing four-week program of instruction.

According to MOI Tashkil Order 1392, released in early April 2013, RTC management became the responsibility of six class "A" province PCOPs in Balkh, Helmand, Herat, Kandahar, Nanghahr,

---

[7] NSOCC-A, "Information Paper on Afghan Local Police," Kabul, Afghanistan, August 9, 2014.

[8] NSOCC-A official, interview with the authors, February 25, 2015.

[9] NSOCC-A official, interview with the authors, February 25, 2015.

[10] NSOCC-A, 2013a.

[11] NSOCC-A, 2013a.

[12] NSOCC-A, 2013a.

and Paktiya.[13] The ALP headquarters training staff supplies the trainers for these facilities and monitors the number of guardians on hand.[14]

The RTC model offers certain advantages and disadvantages. Advantages include standardized training across formations to ensure quality and professionalism, a reduction in the ability of insurgents to conduct insider attacks, and monitoring of access to arms and ammunition.[15] Less tangibly, it is also an advantage to instill in ALP forces a greater sense of connection to GIRoA through involvement in professional training with AUP and ALP from all over the country.[16] Disadvantages of the RTC model include the removal of ALP personnel from the districts they are meant to protect, a lack of funding to transport recruits from their districts to the RTC, and the insecurity that goes along with transporting ALP personnel over large distances. According to Coalition advisers in 2015, the concern expressed by some of their predecessors in 2013 that there would be competition between ALP and other ANP elements for a limited number of RTC trainee slots has not been validated, possibly because of the decision not to expand the size of the ALP force beyond 30,000. As of early 2015, ALP students were not being bumped from scheduled training.[17]

Strategy Execution

The execution of NSOCC-A's RTC plan has gone relatively well, despite some initial setbacks and concerns. As the original model for the ALP regional training concept, the Herat RTC quickly progressed to the stage where AUP instructors were training as many as 100 ALP recruits at a time with minimal oversight from SOF mentors.[18] As described in Chapter Two, however, there were problems in issuing weapons and gear to ALP graduates. According to one Coalition SOF

---

[13] NSOCC-A, "ALP Weekly Update," Kabul, Afghanistan, April 1, 2013d.

[14] NSOCC-A official, interview with the authors, February 25, 2015.

[15] NSOCC-A, 2013a.

[16] NSOCC-A official, interview with the authors, February 23, 2013.

[17] NSOCC-A official, interview with the authors, February 25, 2015.

[18] NSOCC-A official, interview with the authors, February 14, 2013.

adviser, provincial and district officials were reluctant to surrender their authority over the provision of equipment and have prevented guardians from leaving the RTCs with their full kits.[19] At the national level, CJSOTF-A ran into problems in the winter of 2013, when it developed a list of regional ALP training locations without consulting NTM-A and MOI.[20] As it happened, CJSOTF-A's list was in conflict with the installation closure plan that NTM-A and MOI had just negotiated in preparation for the transition. A compromise was reached, however, with the RTCs in question being allowed to remain open until the end of 2014.[21]

Still, in 2013, some Coalition analysts felt that CJSOTF-A's training plan was too optimistic, especially given that the majority of the 13 ALP-producing RTCs would be closed by 2014. If one assumed an ALP training rate at each RTC of 50 per month, the six enduring RTCs (those open after 2014) would not be able to train enough ALP forces to reach the then-stated goal of 45,000 guardians while also keeping pace with attrition. Indeed, they recommended that 12 or 13 RTCs remain open until 2018 in order to meet the expanded ALP tashkil authorization.[22] Yet the Coalition analysts admitted that even this number might not be enough, because neither the Coalition nor MOI had done a training capacity and prioritization analysis for the entire ANP,[23] including AUP, which had a backlog of 20,000 officers who were not school trained.[24]

Lastly, it was unclear in 2013 what the overall costs of training ALP guardians at RTCs would be, who would pay these costs, and how they would pay them. According to one Coalition adviser, ALP per diem and travel expenses associated with training at the RTC were not included

---

[19] NSOCC-A official, interview with the authors, February 14, 2013.

[20] NSOCC-A official, interview with the authors, February 15, 2013.

[21] NSOCC-A official, interview with the authors, February 14, 2013.

[22] NSOCC-A official, interview with the authors, February 25, 2013.

[23] NSOCC-A official, interview with the authors, February 14, 2013.

[24] NTM-A official, interview with the authors, February 17, 2013.

in MOI's budget.[25] In addition, no one had calculated the additional cost of securing the RTCs during periods of ALP training, and a Coalition SOF official doubted whether NSOCC-A would bear these costs.[26] Given that the PCOPs were in charge of the RTCs, he speculated, funding for ALP training, accommodation, and security at the RTC would come from their budgets, while transportation costs would likely be borne by the DCOPs who commanded the ALP trainees.[27]

According to ALP SOAG officials interviewed in 2015, although the RTCs have sufficient operating resources, one continuing issue has been the demand by some of these facilities for extra funding to feed ALP students. The ALP SOAG attributed this to the knowledge that ALP personnel receive a U.S.-funded food allowance once they are on the payroll.[28]

Trainers

AUP has assumed complete responsibility for conducting the program of instruction at the RTCs. The benefits of AUP training, according to Coalition officials, include instilling the rule of law in their ALP charges, as well as strengthening the bond of trust between the two forces.[29] Although they currently have limited visibility over ALP operations in rural areas, Coalition advisers believe that AUP has generally not assumed the Coalition SOF role of providing ongoing training at the district level.[30]

Embedding ANA Special Forces in ALP sites for both security and training purposes was an option touted by SOF advisers in 2013. In the words of one such adviser, the combination of ANA Special Forces and ALP forces was "the only tool that denies safe haven" to the

[25] NTM-A official, interview with the authors, February 14, 2013.

[26] NSOCC-A official, interview with the authors, February 14, 2013.

[27] NSOCC-A official, interview with the authors, February 14, 2013.

[28] NSOCC-A official, interview with the authors, February 25, 2015.

[29] NSOCC-A officials, interviews with the authors, February 14 and 23, 2013.

[30] NSOCC-A official, interview with the authors, February 25, 2015.

insurgents.[31] As mentioned earlier, ANA Special Forces members have occasionally trained ALP guardians in their districts, but NSOCC-A advisers have not received feedback on either the effectiveness of the training or the attitudes of the ANA and ALP forces toward this approach.[32] And even supporters of the ANA Special Forces training mission admitted in 2013 that the commander of ANA Special Operations Command had not been a steadfast supporter of ALP, favoring a direct action role for his special forces.[33]

Training Curriculum

For several years following the inception of the program, the ALP curriculum was a standardized 21-day program consisting of classes on such topics as law (constitution, penal code, use of force), police procedures (handcuffing and searching, defensive tactics, investigations), small-units tactics (smalls arms, ambush tactics, hasty defense), protective topics (improvised explosive device and mine recognition, first aid, identification of friendly forces), and basic skills (marksmanship, physical training). In 2014, a fourth week of training was added to the standard curriculum to allow additional marksmanship training, as well as classes on such topics as human rights and community-oriented policing. This was done in part to improve the international community's perception of ALP.[34]

Although the ALP establishment procedures addressed the issue of additional specialized training in military occupational specialties,[35] Coalition advisers have disagreed about whether it was worth trying to professionalize ALP. To some, ALP recruits should be trained in one of several military specialties to reduce their reliance on AUP. One previous proposal, echoing a 2012 MOI request for more-specialized

---

[31] NTM-A official, interview with the authors, February 22, 2013.

[32] NSOCC-A official, interview with the authors, February 25, 2015.

[33] NTM-A official, interview with the authors, February 22, 2013.

[34] NSOCC-A official, interview with the authors, February 25, 2015.

[35] Islamic Republic of Afghanistan, 2013a.

ALP training, recommended focusing on five specialty categories: logistics, maintenance, bomb detection, communications, and medical.[36] But the need for expanded training was not universally recognized. A Coalition adviser observed in 2015 that AUP was supposed to manage all the support functions for ALP and that other forces were supposed to provide Quick Reaction Forces when necessary. Therefore, according to one NSOCC-A official, "once you train on the basics and checkpoints, that's all you need."[37] Despite the recent increase to a four-week program of instruction, this opinion mirrors what has historically been the Coalition's response to initiatives for increased training, presumably because of their uncertain cost and effectiveness, threat to higher-priority initiatives, and potential to remove the onus of supporting ALP from MOI.[38]

Leaving aside the question of whether increased ALP professionalization is cost-effective, it is clear that ALP guardians receive less logistics training than other elements of ANP.[39] The ALP program of instruction in 2013 contained only two of the seven logistics classes offered by MOI,[40] and guardians were not trained at all on the topics of equipment replacement and sustainment.[41] Training slots are controlled by the respective PCOPs and DCOPs, and Coalition advisers reported in 2015 that they were not inclined to send ALP personnel to advanced courses; the advisers also stated that, given AUP's responsibility for logistics and sustainment, it did not appear that ALP would be given such training in the future.[42] Furthermore, while some ALP units have had basic triage training with the Red Cross,[43] the lack of

---

[36] MOI officials, interviews with the authors, February 21, 2013; and NTM-A officials, interviews with the authors, February 17 and 21, 2013.

[37] NSOCC-A official, interview with the authors, February 25, 2015.

[38] NTM-A official, interview with the authors, February 17, 2013.

[39] Regional Command East official, interview with the authors, February 13, 2013.

[40] MOI official, interview with the authors, February 19, 2013.

[41] NTM-A official, interview with the authors, February 17, 2013.

[42] NSOCC-A official, interview with the authors, February 25, 2015.

[43] SOF team member, interview with the authors, February 2013.

ALP medical training, in addition to the lack of training in leadership and operational planning, has been a problem.[44]

Mobile Training Teams

In addition to the occasional assignment of ANA Special Forces members to conduct ALP training in the districts, MOI has formed mobile training teams that are available to travel to the ALP units. This is similar to a model that has been used with success in the Ministry of Defense in training ANA Special Operations Command kandak forces that are unable to leave their assigned area. If ALP units cannot travel to a training center because of security, manpower, or logistics reasons, the mobile training team provides a means to replicate some of the material covered in the RTC program of instruction. In addition to offering basic training, these mobile teams can provide instruction on specialty topics, such as advanced first aid and countering improvised explosive devices.[45] As of February 2015, however, mobile training teams had not been extensively employed, with one Coalition adviser approximating that they conducted only 5 percent of training events.[46] But considering the extent to which these mobile training teams are employed, it is conceivable that MOI officials would take note and employ them to solve some of their security and transportation problems in the future.

Conclusion

The state of ALP training is good when compared with the rest of the Afghan police force. Nearly all of the ALP guardians on the books in early 2015 had received their required four weeks of training. However, there were still security and logistics concerns in transporting ALP

[44] Regional Command East official, interview with the authors, February 13, 2013.

[45] NSOCC-A official, interview with the authors, February 25, 2015.

[46] NSOCC-A official, interview with the authors, February 25, 2015.

guardians to training centers, and these concerns must be addressed by PCOPs and DCOPs, who may have other priorities. Thus, most Coalition advisers agreed that a hybrid training system—with local and regional aspects, and possibly the use of mobile training teams—was the best option for the future.

CHAPTER SIX

Recommendations for Future Local Security Programs

In cooperation with MOI and CSTC-A, NSOCC-A is undertaking to institutionalize the ALP program and help build the capacity of MOI to support ALP guardians at the village level. This effort, which builds on cooperative endeavors over the past several years, addresses many of the functional and cross-cutting issues that hamper the development and sustainability of ALP. While some institutionalization initiatives have fared better than others and some are still a work in progress, we have seen positive results—in the form of recruits processed and vetted, equipment received, guardians paid, and candidates trained—with limited Coalition assistance since the end of 2014. However, examples of improved capabilities and attitudes on the part of some MOI and ALP officials in Kabul and in some provinces and districts do not yet equate to an institutional commitment and capacity to sustain the ALP program. Furthermore, because of the Coalition's withdrawal from most rural parts of the country and almost total reliance on Afghan reporting, it is no longer possible for NSOCC-A to systematically and reliably gauge the level of MOI support to ALP where it matters most—in the villages where the guardians reside.

Still, for future U.S. government efforts to assist foreign partners in building the capacity of their local security forces, there are important lessons that can be learned from the experience of transitioning ALP to full Afghan control and attempting to make it a key element of the national police force.

Lesson 1: Advisers Must Take Account of the Operating Environment and Work in Concert with Various Partners

Chapter Two described some of the obstacles extending across the MOI functions that support the ALP program. For example, the lack of MOI and ALP headquarters authority over PCOPs makes it difficult to ensure that local administrators are responding to the logistics, personnel management, and training needs of ALP units, as well as making those units accountable for the resources they receive. Additional challenges are created by the uncertain commitment of senior MOI leaders to the ALP program, the difficulties in assessing conditions in ALP districts given poor communications, irregular transportation and unreliable data systems, and the low level of education among ALP personnel. Although U.S. advisers are unlikely to confront the same array of issues in the future as they did in Afghanistan, they could face similar obstacles when assisting governments in poor, conflict-torn countries with limited infrastructure and human capital. In such cases, they must do their best to first understand the lay of the land and then recommend a support plan that either circumvents or erodes potential blockages. This happened to some extent in Afghanistan. When it became clear that ALP forces in many locations had not received their allotted equipment, NSOCC-A and NTM-A advisers worked with ALP headquarters and local police officials to push initial ALP supply packages to districts throughout the country. Furthermore, recognizing that this supply method was not sustainable without continuing U.S. assistance, they persuaded the Afghans to switch to a system that provided individual equipment to ALP trainees at RLCs. However, the political, administrative, and technical hurdles associated with developing a pull-based resupply process for ALP have not been overcome, despite significant U.S. efforts.

Furthermore, U.S. advocates have not completely surmounted the chief roadblock to MOI's providing adequate support to ALP—that is, the resistance to the program on the part of officials in the Afghan police and, to a lesser extent, within U.S. and international aid bureaucracies. Other reports on lessons learned from capacity-building have stressed the importance of working on security assistance plans with officials in the host nation.[1] Moreover, in Afghanistan, the decision by Coalition SOF to build a local security force from the ground up largely on its own—albeit with the enthusiastic support of the Coalition force commander—had some positive consequences. SOF's commitment to and experience in conducting village stability operations gave the ALP program a momentum that it might otherwise have lacked, as well as a set of guide posts that kept the program largely on course. On the other hand, the proprietary nature of the ALP enterprise made it challenging to transition the program to MOI as the Coalition began its withdrawal or to obtain the enthusiastic backing of Coalition agencies responsible for advising and assisting the bulk of ANP (of which ALP was a relatively small contingent). Although NSOCC-A officials eventually recognized the need to integrate ALP into AUP and worked hard to convince their Afghan and Coalition counterparts of the program's worthiness, their outreach efforts came rather late in the game and were perceived by some as half-hearted.

Lesson 2: Pull-Based Logistics Systems Often Take a Long Time to Evolve

As just noted, the Afghans have so far maintained the Coalition's proposed system for supplying new guardians with their initial kits, but the resupply situation has been much more problematic. Thus, NSOCC-A and its Coalition partners have faced a quandary throughout ALP's existence: Should they continue to push as much equipment and supplies as authorized to ALP districts (to avoid their being overrun by the Taliban) or insist that MOI provide an accurate accounting of existing inventories of ALP supplies before releasing additional resources (out of a suspicion that a significant portion would not reach their intended destination)? That this dilemma has never been satisfactorily resolved has much to do with the length of time it takes to establish an effective

[1] See, for example, Christopher Paul, Colin P. Clarke, Beth Grill, Stephanie Young, Jennifer D. P. Moroney, Joe Hogler, and Christine Leah, *What Works Best When Building Partner Capacity and Under What Circumstances?* Santa Monica, Calif.: RAND Corporation, MG-1253/1-OSD, 2013.

pull-based stock replenishment system in a place like Afghanistan, if it can be done at all.

In Chapter Three, we observed that MOI's equipment distribution process worked relatively well from MOI's national warehouses to the RLCs. But the process became murky and uneven when equipment left the regional warehouses bound for provincial and district headquarters. At that point, the decision of what supplies and how much the ALP guardians received was at the discretion of PCOPs and DCOPs, some of whom appreciated and supported ALP forces and others of whom either did not support them or had higher equipping priorities than ALP. The result was a fair deal of hoarding and diversion of resources at various levels. Yet the total amount was not at all clear because of the lack of accountability below the RLC. For their part, Coalition officials insisted that they had the paperwork for the initial ALP equipment issue and would not provide replacement items unless the Afghans could document what had happened to the old equipment. Furthermore, they contended, ALP headquarters officials were learning how to audit, and had had an opportunity to demonstrate their proficiency during staff assistance visits to the provinces. But the officials also acknowledged that Coalition advisers did not have the ability to accompany these auditing teams.

What does this experience suggest about what needs to be done to sustain U.S. military capacity-building initiatives in other parts of the world? On the one hand, it indicates that, in situations where the existing supply chain is long, broken, or undeveloped, U.S. advisers and partner government officials may have little choice initially but to push resources to units in the field to ensure that they have the wherewithal to defend themselves. On the other hand, they should begin as soon as possible to put in place a supply system that allows for positive control through property accountability or inventory records so that equipment and supplies can be tracked and their organizational owners held responsible for their maintenance and use—for the sake of both operational effectiveness and the transparency necessary to ensure long-term political support for security assistance missions. Of course, this second step will not be accomplished easily. As three experts in defense institution reform noted, establishing a mature logistics system is "an evolutionary journey."[2] Rather than attempting to make the immediate leap to a first-tier, pull-based stock replenishment system, donors should consider simpler alternatives that account for the partner's level of resources, literacy, technical competence, communications, and data availability. For example, recommendations on information systems and methods should be based on technology in broad use within the partner country so that they can be maintained at an affordable cost without continual external assistance. In addition, advisers and partner officials should look for creative ways to collect and analyze available information that might contribute to a greater understanding of logistics requirements and consumption patterns. In the longer term, they should invest in logistics-focused human capital development so that high-quality personnel can be recruited and trained, thereby setting the stage for a higher-functioning logistics operation.

Lesson 3: Managing Dispersed Forces Requires a Balance Between Local Autonomy and Central Oversight

In Chapter Four, we concluded that, despite signs that not all ALP members are being properly recruited and paid, MOI is beginning to develop effective processes for the administration of ALP personnel. Positive signs include the spread of electronic personnel payment methods to reduce the corruption and delay associated with the traditional "trusted agent" system, as well as the establishment of indigenous mechanisms for vetting and in-processing ALP recruits. Nevertheless, managing ALP personnel in widely dispersed locations has been challenging. This is partly attributable to transportation and communication hurdles, partly to the paucity of qualified local administrators, and partly to the exigencies of Afghanistan's political patronage system and ethnic and tribal differences. As a result, provincial and district police officials have been given a relatively free hand to support, neglect, or exploit ALP forces. While acknowledging that a completely centralized

---

[2] Unpublished 2014 RAND research on building a logistics system within defense organizations, by Eric Peltz, Frederick M. Boomer, and George Topic.

managerial structure is not possible under Afghan conditions, Coalition advisers contend that MOI's decentralized command and control structure has constrained those at the top from communicating with those at the bottom and gaining insight into what is happening at lower levels, as well as from pressuring mid-level commanders into taking effective action to meet the requirements of ALP guardians in the villages.

What this suggests for future capacity-building initiatives is the need to find a balance between encouraging local leaders to take charge of the daily management of local security forces and ensuring that the former raise and employ the latter appropriately and continue to provide adequate support to them. Although it is ultimately up to the host-nation government to determine appropriate managerial structures, policies, and processes, external advisers should do what they can to help host-nation leaders overcome handicaps to effective oversight. For example, they could advise the national headquarters to improve its downward lines of communication and influence by placing trusted agents at RLCs and local administrative headquarters. Beyond that, they could back efforts to place local police officials under national ministerial authority in places where they currently are not, such as Afghanistan. That said, interfering with a country's governance structures could create unwanted legal and political complications.

Lesson 4: Centralized Training Has Advantages, but a Hybrid System May Work Best Over the Long Term

As Chapter Five indicated, the state of ALP training appeared to be relatively good in the winter of 2015, with nearly all of the registered ALP guardians receiving their required four weeks of training. Still, there was some concern expressed regarding the capability and willingness of local police officials in insecure or far-flung districts to continue transporting ALP recruits to RTCs. In addition, despite its success in turning out a steady supply of ALP graduates, regional training is not a substitute for the localized training that SOF teams provided in villages throughout much of Afghanistan from 2010 to 2014. Although centralizing the initial training of ALP forces improves the chances that the Afghans can sustain the program, follow-on local training is still useful to maintain and improve ALP skills, as well as to establish bonds of mutual support and trust between village guardians and other security forces in their vicinity. It also provides an opportunity for training providers to assess and monitor local security force performance in the field.

In looking to the future, local security force advisers should consider recommending a hybrid (local and regional) training system to partner government officials. This would require two planning steps. The first is a comprehensive assessment of the training needs of all of the elements of the police force, as well as the capacity of RTCs to meet these needs. Second, based on the results of this assessment and a detailed evaluation of what is realistic and appropriate for particular elements of the local security force, external advisers and host-nation officials should develop training plans that employ a combination of RTCs, local training venues, and mobile training teams. As part of the training assessment, advisers and officials should review the program of instruction for local security recruits for gaps and consider more-advanced skills training (e.g., medical, logistics, and communications training) for leaders and more-qualified village guardians. Of course, this will be challenging in cases, like Afghanistan, where even village leaders lack basic literacy skills. Furthermore, additional training requires more funding, both for instruction and possibly for higher salaries for those trained. One of the appeals of local security forces is their relatively low cost. Also, bolstering the support capabilities of ALP or similar forces may take the onus off the overall police organization for caring for those at the bottom of the chain. However, the risk in not providing adequate training in support skills is that local security forces will be ill-prepared to stand on their own or garner higher-level assistance once external providers depart.

Lesson 5: The Coalition Advisory Structure Should Be Maintained Until the Host Nation Has an Assured Sustainment Capability

Despite promising developments in logistics, personnel management, and training, it is still unclear whether Afghanistan's MOI has the will and capability to independently support the ALP program without outside assistance and oversight. Ideally, the Coalition would have continued to embed advisers at MOI headquarters, regional logistics and training centers, and provincial and district headquarters for several years following the transition of ALP support responsibilities to the host-nation government—and this was NSOCC-A's original plan. The relatively quick turnover of control to MOI and concentration of the advisory function at the national level—without a reliable means of gauging support requirements and receipts in the field—jeopardize the success that ALP guardians have achieved to this point and make the program vulnerable to criticism from those who question whether U.S. government dollars are being wisely spent. This is true for the following reasons. First, MOI is still a developing ministry whose commitment to the ALP program, and whose capacity to supervise and support it, requires bolstering by the Coalition. Second, provincial- and district-level support for ALP is highly uneven, and PCOP and DCOP links to MOI and ALP headquarters are often problematic.

Thus, if the United States and its allies desire to help build local security forces in the future in countries similar to Afghanistan, with limited infrastructure and weak governance, they should consider maintaining an advisory presence in key regions, or some type of circuit rider or distributed operations scheme, in the period following the transition. This would permit coalition advisers to continue to work with headquarters officials and local leaders on resolving resource, personnel, and training management issues pertaining to the police and military. Of course, such a plan has financial and security implications for external providers of assistance. Most importantly, advisers working or traveling in the field would need a reasonable level of force protection, as well as the assurance that they would be transported expeditiously to a modern medical facility if they were injured. This is not to say that outsiders should assume the host government's role of ensuring adequate support to local security forces. External advisers should focus on encouraging local officials to establish their own mechanisms for managing the support requirements of local security forces, as well as overseeing the implementation of these managerial functions.

Maintaining a mentoring and oversight presence that can interact with all levels of the relevant local ministries is a necessary but insufficient condition to ensure that the host-nation government can support local security forces. To achieve this goal, external advisers should begin testing the local government's ability to independently perform the gamut of logistics, personnel management, and training functions as soon as it is feasible to do so. To an extent, NSOCC-A did this in ALP districts where SOF teams ended their training mission by moving into tactical overwatch positions; from these positions, they could monitor the performance of local ALP units and their AUP command structure while remaining prepared to step in to assist them in emergency situations. However, Coalition SOF continued to play an essential role in meeting ALP logistics and pay needs up until the point of transition. And then when transition to tactical overwatch occurred, the process ensued rapidly and not always with sufficient consideration given to district-level conditions or SOF's ability to observe and advise local ALP and AUP officials. With respect to similar capacity-building missions in the future, the United States and its allies should begin planning to turn over support functions to the host nation from the start, so they can evaluate the results of the turnover and help fix any problems that might exist prior to their departure from the field.

References

Asia Foundation, *A Survey of the Afghan People*, Washington, D.C., November 17, 2015.

Curtis, Ashley, "Logistics Shura Encourages ANP Self-Sufficiency," Kandahar Airfield, Afghanistan, 117th Mobile Public Affairs Detachment, June 10, 2012.

Deputy Commander of Special Operations Forces, NTM-A, "Info Paper 12-05," Kabul, Afghanistan, 2012b.

———, "Primer: Mission Analysis/Needs Analysis," Kabul, Afghanistan, 2012c.

Fioriti, Joris, "Local Police, an Uncertain Player in Afghan Future," Goshta, Afghanistan: Agence France-Presse, December 31, 2012.

Helmus, Todd C., *Advising the Command: Best Practices from the Special Operations Advisory Experience in Afghanistan*, Santa Monica, Calif.: RAND Corporation, RR-949-OSD, 2015. As of January 21, 2016:
http://www.rand.org/pubs/research_reports/RR949.html

Hewad, Gran, "The 2015 Insurgency in the North (4): Surrounding the Cities in Baghlan," Afghanistan Analysts Network, October 21, 2015. As of October 21, 2015:
https://www.afghanistan-analysts.org/insurgency-in-the-north-4-baghlan/

Human Rights Watch, "Afghanistan: Rein in Abusive Militias and Afghan Local Police," Kabul, Afghanistan, September 12, 2011. As of October 26, 2015:
https://www.hrw.org/news/2011/09/12/
afghanistan-rein-abusive-militias-and-afghan-local-police

International Crisis Group, *The Future of the Afghan Local Police*, Brussels, Belgium, Asia Report No. 268, June 4, 2015. As of October 28, 2015:
http://www.crisisgroup.org/~/media/Files/asia/south-asia/pakistan/
268-the-future-of-the-afghan-local-police.pdf

International Security Assistance Force, "Personnel and Payroll System Relationships," presentation, Kabul, Afghanistan, December 30, 2014.

Islamic Republic of Afghanistan, *Ten-Year Vision for the Afghan National Police: 1392–1402*, Kabul, Afghanistan: Ministry of Interior Affairs, 2013a. As of January 21, 2016:
https://ipcb.files.wordpress.com/2013/06/
13-04-02-ten-year-vision-english-final-version.pdf

———, *ALP Policy and Establishment Procedures*, Kabul, Afghanistan: Ministry of Interior, April 2013b.

Jones, Seth, G., *The Strategic Logic of Militia*, Santa Monica, Calif.: RAND Corporation, WR-913-SOCOM, 2012. As of January 21, 2016:
http://www.rand.org/pubs/working_papers/WR913.html

Madden, Dan, "The Evolution of Precision Counterinsurgency: A History of Village Stability Operations & the Afghan Local Police: Commander's Initiative Group," CFSOCC-A Commander's Initiative Group, June 30, 2011.

Moyar, Mark, *Village Stability Operations and the Afghan Local Police*, Joint Special Operations University, October 2014.

Nadir, Ali, "SOF, Shuras, and Shadow Governments: Informal Governance and Village Stability Operations," CFSOCC-A Commander's Initiative Group, 2011.

NATO—*See* North Atlantic Treaty Organization.

NATO Special Operations Component Command–Afghanistan, "ALP Weekly Update," Kabul, Afghanistan, January 21, 2013a.

———, "Information Paper, 20 Feb 2013," CJ4, Kabul, Afghanistan, 2013b.

———, "ALP Weekly Update," Kabul, Afghanistan, March 1, 2013c.

———, "ALP Weekly Update," Kabul, Afghanistan, April 1, 2013d.

———, "ALP Weekly Update," Kabul, Afghanistan, April 8, 2013e.

———, "ALP Weekly Update," Kabul, Afghanistan, April 15, 2013f.

———, "ALP Weekly Update," Kabul, Afghanistan, April 22, 2013g.

———, *Afghan Local Police In-Processing Enrollment Handbook*, Kabul, Afghanistan, May 2013h.

———, *Afghan Local Police Pay and Logistics Guide*, Kabul, Afghanistan, May 2013i.

———, "ALP Weekly Update," Kabul, Afghanistan, May 20, 2013j.

———, "Information Paper on Afghan Local Police," Kabul, Afghanistan, August 9, 2014.

———, "ALP Strength Report," Kabul, Afghanistan, February 21, 2015a.

———, "ALP Introductory Brief," Kabul, Afghanistan, February 25, 2015b.

New America Foundation, "Attacks on U.S. and NATO Soldiers by Afghan Security Forces," web page, October 18, 2012. As of May 6, 2013: http://newamericafoundation.github.io/security/maps/afghanistan.html

North Atlantic Treaty Organization, *NATO Logistics Handbook*, Brussels, Belgium, October 1997. As of January 21, 2016: http://www.nato.int/docu/logi-en/logist97.htm

NSOCC-A—*See* NATO Special Operations Component Command–Afghanistan.

Parrish, Karen, "Special Ops Task Force Helps Shift Afghanistan Trend Line," American Forces Press Service, May 13, 2013. As of January 21, 2016: http://archive.defense.gov/news/newsarticle.aspx?id=120051

Paul, Christopher, Colin P. Clarke, Beth Grill, Stephanie Young, Jennifer D. P. Moroney, Joe Hogler, and Christine Leah, *What Works Best When Building Partner Capacity and Under What Circumstances?* Santa Monica, Calif.: RAND Corporation, MG-1253/1-OSD, 2013. As of January 21, 2016: http://www.rand.org/pubs/monographs/MG1253z1.html

Saum-Manning, Lisa, VSO/ALP: Comparing Past and Current Challenges to Afghan Local Defense, Santa Monica, Calif.: RAND Corporation, WR-936, 2012. As of January 21, 2016: http://www.rand.org/pubs/working_papers/WR936.html

Special Operations Joint Task Force–Afghanistan, "ALP Special Operations Advisory Group," policy memorandum, Kabul, Afghanistan, February 26, 2015.

SOJTF-A—*See* Special Operations Joint Task Force–Afghanistan.

U.S. Department of Defense, *Report on Progress Toward Security and Stability in Afghanistan*, Washington, D.C., November 2010. As of January 21, 2016: http://archive.defense.gov/pubs/November_1230_Report_FINAL.pdf

U.S. Department of Defense Inspector General, *Special Plans and Operations: Assessment of the U.S. Government and Coalition Efforts to Develop the Afghan Local Police*, Alexandria, Va., July 9, 2012.

Special Inspector General for Afghanistan Reconstruction, *Quarterly Report to the United States Congress*, Arlington, Va., April 30, 2015a.

———, *Afghan Local Police: A Critical Rural Security Initiative Lacks Adequate Logistics Support, Oversight, and Direction*, Arlington, Va., SIGAR 16-3 Audit Report, October 2015b. As of January 21, 2016: https://www.sigar.mil/pdf/audits/SIGAR-16-3-AR.pdf